Ground Squirrel
Front L 1.6 W 1.3
Hind L 2.5 W 1.9

Prairie Dog
Front L 2.5 W 2.0
Hind L 2.5 W 2.3

Eastern Chipmunk
Front L 2.2–2.5 W 0.95–2.1
Hind L 1.3–1.9 W 1.6–2.4

Gray Squirrel
Front L 4.1 W 2.5
Hind L 6.6 W 2.5–4.4

Red Squirrel
Front L 3.0 W 1.9
Hind L 4.4–6.3 W 4.4–7.3

Kangaroo Rat
Front L 2.5 W 2.8
Hind L 3.8 W 1.3

Deer Mouse
Front L 0.63 W 0.63
Hind L 0.83 W 1.3

Jumping Mouse
Front L 1.0 W 0.63
Hind L 2.5 W 0.63

North American Porcupine
Front L 5.7–8.4 W 3.2–4.8
Hind L 6.3–9.8 W 3.8–5.1

Cottontail Rabbit
Front L 2.5 W 1.6
Hind L 7.6 W 2.5

Snowshoe Hare
Front L 4.4 W 3.8
Hind L 11.4–12.7 W 9.5–11.4

Virginia Opossum
Front L 3.8–5.4 W 4.4–6.0
Hind L 4.4–6.3 W 4.4–7.3

Nine-banded Armadillo
Front L–4.6 W 3.6
Hind L 5.6 W 4.1

***Sorex* Shrews**
Front L 0.63 W 0.51
Hind L 0.76 W 0.51

Artists: Susan C. Morse and Jesse Guertin

ARTISTS' CREDITS
Sandra Doyle/Wildlife Art Ltd.: Plates
  86–89, 97–110
Nancy Halliday: Plates 2–8, 26–30
Ron Klingner: Plates 9–13, 43–47
Elizabeth McClelland: Plates 31–42,
  90–96
Consie Powell: Plates 74–85
Wendy Smith: Plates 48–53, 63–73
Todd Zalewski: Plates 1, 14–25, 54–62
Cover illustration: Todd Zalewski
Scat illustrations: Diane Gibbons
Track illustrations: Susan C. Morse and
  Jesse Guertin

Roland W. Kays is the Curator of
Mammals at the New York State Museum.
His research centers on the ecology and
conservation of temperate and tropical
mammals, especially carnivores.
Don E. Wilson is Senior Scientist at the
Smithsonian Institution's National
Museum of Natural History. Recipient of
a Smithsonian Institution Award for
Excellence in Tropical Biology and a U.S.
Fish and Wildlife Service Outstanding
Publication Award, he is the author or
coauthor of more than 180 scientific
papers and twelve books, including three
volumes on bats, *The Smithsonian Book of
North American Mammals*, and *Mammal
Species of the World*.

Copyright © 2002 by Princeton
  University Press
Published by Princeton University Press
41 William Street
Princeton
New Jersey 08540

In the United Kingdom:
Princeton University Press
3 Market Place
Woodstock
Oxfordshire OX20 1SY

Library of Congress Cataloging-in-Publication Data
Kays, Roland, 1971-
    Mammals of North America/Roland Kays and Don E. Wilson.
      p. cm.—(Princeton field guides)
    Includes bibliographical references (p.).
    ISBN 0-691-08890-X (cl : alk. paper)—ISBN 0-691-07012-1 (pb : alk. paper)
    1. Mammals—North America—Identification. I. Wilson, Don E. II. Title. III. Series.
QL715 .K38 2002
599'.097—dc21   2002024305

British Library Cataloging-in-Publication Data is available

This book has been composed in Galliard (main text) and ITC Franklin Gothic
(headings and tabular material)

Printed on acid-free paper

www.pupress.princeton.edu

Edited and designed by D & N Publishing, Marlborough, Wiltshire, UK

Printed in Italy

9 8 7 6 5 4 3 2 1

CONTENTS

ACKNOWLEDGMENTS

The trick to writing a good field guide is to collect the scattered information on species identification and concentrate it into one concise volume. As a whole, North American mammals are a well-studied group, and the details in this book are largely a credit to the work produced by generations of mammalogists. Although books and journal articles provided some of this detail, no field guide could be completed without substantial research in natural history museums, and this guide is no exception. Our own institutions, the New York State Museum and the National Museum of Natural History, provided the collections, library facilities, and a wide variety of witting and unwitting ancillary support. In addition, we thank the Denver Museum of Nature and Science, the University of Colorado Museum, and the Field Museum, for access to their collections. A specimen in the drawer is much easier to bring to life when an expert helps point out the relevant characters. For this museum and identification help we thank:

Greg Anderson, Andrea Bixler, Joe Bopp, Harold Broadbooks, Mike Carleton, Linda Gordon, Lawrence R. Heaney, Al Hicks, Rosanne Humphrey, Cheri A. Jones, Zack Knight, Bill Longland, Chris Maser, Jesus Maldonado, James Mead, Bruce Patterson, Toni Piaggio, John Phelps, Roger A. Powell, Eric Rickart, Dave Schmidt, Michi Schulenberg, Bob Smith, William Stanley, and Westarp Wissenschaften. Additionally, we thank William Gannon for his assistance with chipmunk vocalizations, Dan Simberloff for his help with introduced species, and Al Hicks for plunging into caves to help us get the bats just right. Tim Page carefully read and corrected the entire text. A special thanks to Wade Sherbrooke and the staff at the Southwest Research Station for their hospitality during one of our field trips. We thank Mike Carleton and Guy Musser for providing common names from their muroid rodent section in the next edition of *Mammal Species of the World* (Wilson and Reeder, 1993).

The range maps were provided in a GIS for this field guide by Wes Sechrest. In collaboration with many conservation, museum, and academic groups including Conservation International, the Institute of Applied Ecology, and the International Union for the Conservation of Nature (IUCN), Sechrest has compiled distribution information for all mammal species in the world. The initial products, extent of occurrence maps for all mammal species, will soon be freely available on the new Species Information Service (SIS) of the IUCN.

A field guide is part science and part art. For help and critique of artwork we thank:

Catherine Chapman, Marlene Hill Donnelly, Ben Flemer, Patricia Kernan, Clara Richardson Simpson, David Steadman, and Karen Teramura, the staff at WILDlifeART, and Christa Wurm. Thanks to Keeping Track's (www.keepingtrackinc.org) founder and Program Director Susan Morse for taking the time off their citizen-based wildlife monitoring, educational, and conservation programs to provide her expertise regarding the animal tracks and scat illustrations.

We thank Greg Anderson, Tom Brooks, Chris Byrne, Chip Foster, Fritz Hertel, Barret Klein, Darrin Lunde, Chris Skelton, and John Young for insightful discussions about field guide design. Thanks to Ron Gill, Adam Fox, and Dimitri Karetnikov for computer help. Our editors at Princeton University Press, Sam Elworthy and Robert Kirk, gently pushed and pulled at appropriate points throughout the process, contributing greatly to the timely completion of the book.

While the above offered their "perspiration," we owe thanks to another suite of individuals for their inspiration and encouragement, especially Jim Findley, John Gittleman, Bonnie and David Kays, Deedra McClearn, Kevin McGowan, Karen Zich Reiss, and Bruce Patterson. Most importantly, we thank our wives Judy Kays and Kate Wilson, for their continuing graceful tolerance of our long hours in the field and museum over the years.

INTRODUCTION

A Moose crosses the road and traffic stops. Fearful campers scrutinize a Black Bear as it moves through their campsite. Even a mouse scurrying across the ground in the back yard will attract notice. Mammals command attention. Except for a few common species, most of our 442 mammal species are rarely seen; when they do show themselves, it can be quite exciting.

Mammals arouse our emotions, often in contradictory ways. The fluffy Gray Squirrel is awfully cute, until it nests in your attic. The charismatic Gray Wolf is a majestic symbol of wilderness, until it threatens your livestock. Depending on your point of view, a White-tailed Deer is a precious little Bambi, a trophy to be mounted on the wall, a hunk of meat to be sizzling on the grill, a pest that won't leave your garden alone, or a 300-pound roadblock that could jump in front of your car at any moment.

These anthropomorphic views are obvious, but the less obvious ecological duties fulfilled by our mammals may be more important. The diverse ways in which they make a living means that they play myriad ecological roles that are at the very core of a healthy environment. Granivorous mammals (e.g. squirrels and mice) eat seeds, killing many potential plant offspring; but they also disperse some seeds away from the shade of the mother plant unharmed and into a good environment for germination. Thus they sometimes act as a friend of the plant, sometimes as a foe. Folivorous mammals (e.g. deer and rabbits) eat the leaves of plants and can keep certain species from overgrowing an area. Carnivorous mammals (e.g. weasels, Bobcats, and Killer Whales) keep their prey populations in check by eating the most abundant species. Predators can actually increase prey diversity by preventing a single species from becoming overabundant and driving others out. Insectivorous mammals (e.g. bats and shrews) help control insect populations, including many pest species. Our diverse and abundant mammal fauna constitute an important, well integrated part of our varied ecosystems.

There are a number of reasons why one might want to identify a mammal, from idle curiosity to hard-core scientific inquiry. Anyone with a nose for nature will be curious about mammals encountered in the wild. The relative rarity of those encounters makes them all the more special, and our "biophilia," or attraction to animals, carries with it a strong desire to identify whatever we see. For most of us, this book will serve as a handy guide in pocket or vehicle, for casual encounters with mammals. For others, it may be a useful companion on field trips or research projects designed to find and identify specific mammal species.

SPECIES INCLUDED

This field guide is designed to efficiently and accurately identify all 442 mammal species known from North America north of Mexico. In addition to native species, this includes tropical species that rarely venture north across the border, exotic species introduced from foreign lands, and extinct species.

Some species on our list have very rarely been recorded in North America (e.g. Margay and Hairy-legged Vampire), but deserve coverage in a field guide so that naturalists know what to look for in the event that these species recolonize.

Selecting which introduced mammal species to include in a field guide is a bit tricky. We selected only exotic mammal species known to survive and reproduce in the wild. This includes a number of ungulate and rodent species that could be confused with our native species. We decided not to include species such as the domestic horse or domestic cat, which have feral populations but are known to everyone, and unlikely to be confused with any native mammals. We did include the domestic dog for a comparison with the similar Coyote. Because they are unlikely to be confused with native fauna, we did not include the various monkey species known to have escaped from zoos or breeding centers and that survive in some areas in Florida, and perhaps Texas.

WHAT IS A SPECIES?

Scientists continue to refine our definition of the term species. Traditionally, we recognized independent evolutionary lineages that were reproductively isolated from other such lineages as species. This biological species concept is often difficult to apply objectively. Recent advances in molecular technology that allow much greater resolution of evolutionary relationships have led to a continuing reexamination of our ideas about what constitutes a species. For purposes of this book, we have followed Wilson and Reeder (1993) in determining which species to recognize. A revision of that work, to be published in 2003, was available to us in manuscript form, allowing us to include changes that have occurred in the past decade. So, for our purposes, a species is an evolutionary lineage of mammal that is recognized as distinct from other such lineages by appropriate authorities in the field of mammalogy.

WHAT INFORMATION IS INCLUDED?

We designed this book to be useful both to amateur naturalists and professional mammalogists. We have packed as much detail into this volume as possible, but also worked hard to keep it concise and efficient, so that it is not unwieldy in the field. All the information for a given species is displayed on two facing pages.

ARTWORK

The 108 color plates are the core of this guide and will be the most useful component for mammal identification. Each species is represented by an illustration, and some species have a number of illustrations to show variations in their appearance due to age, sex, season, or geographic variation. All paintings are original works of art created for this book and are the result of close collaboration between the artists and ourselves. Credits for each plate are listed on the copyright page.

A NOTE ON SKELETAL MATERIAL

Most species can be identified based on external appearances. However, sometimes two species of mammal can be distinguished only by a dental or skeletal character. In these cases we include an illustration of such differences.

DISTRIBUTION MAPS

The maps show the most recent geographic distribution data for each species. These are the products of a Herculean effort to map the distribution of every mammal species of the world and are available in electronic format at the IUCN's species information service web page. Mammal ranges are dynamic, with some populations expanding into new areas and others becoming extirpated locally. Readers should therefore be on the lookout for species outside of their normal geographic range.

SPECIES ACCOUNTS

Each species account has a brief paragraph noting a species' common name, scientific name, measurements, and details about appearance, general ecology, and behavior. All species accounts are written with a specific formula to make it easy to find the bit of information you are looking for:

COMMON NAME *Genus species* total length, tail length, weight
(with differences for male and female where significant)

1) The single most important piece of information for identifying the species.
2) Description of the physical traits of the species. Details of within-species variation and how to tell it apart from similar species.
3) Description of behavior, where relevant, including vocalizations.
4) Description of habitats used by a species.

SCAT ILLUSTRATIONS

Mammals are often elusive, but their scat (feces) can be easy to find. This makes scat an excellent tool to document the presence of local mammals, and some mammalogists get quite excited over a good scat discovery. To aid in scat study, we have included examples of mammal scat shapes (*see* pages 230–233). While some species have very characteristic scats, others can be quite variable, with the shape and size dependent on the contents of the animal's last meal. Generally, herbivores have more consistent diet, and therefore their scat can be more reliably identified. Carnivores and omnivores have more variable diets, and therefore are much more difficult to identify by scat alone.

—— MAMMAL TRACKS

Most big or medium-sized mammals have characteristic footprints that can be detected and identified in soft mud or snow. We have provided illustrations of these tracks, with a range of measurements, on the inside flap of the cover of this guide.

WHAT INFORMATION IS NOT INCLUDED?

While this book has all the information you need to identify every mammal species in North America, it does not have much more. To keep this field guide efficient to use, small, and easy to take into the field, we have minimized discussion about the ecology, behavior, evolution, and conservation of each species. We encourage readers to read more about the mammals they see and identify, and recommend the following sources for their bookshelves and coffee tables – but not necessarily for their field backpack.

FURTHER READING

Glass, Brian P., and Monte L. Thies, 1997. *A Key to the Skulls of North American Mammals*, 3rd edn. Oklahoma State University.

Nowak, Ronald M., 1999. *Walker's Mammals of the World*, 6th edn, vols 1 & 2. Johns Hopkins University Press.

Wilson, Don, and Sue Ruff (eds), 1999. *The Smithsonian Book of North American Mammals*. Smithsonian Institution Press, Washington.

Wilson, Don E., and DeeAnn M. Reeder (eds), 1993, *Mammal Species of the World*. Smithsonian Institution Press.

Regional Tracking guides, by J. Halfpenny, published by Pequot Press.

The Journal of Mammalogy, published 4 times a year. Hard-core science, not for the faint of heart.

RECOMMENDED INTERNET RESOURCES

- The Animal Diversity Web: http://animaldiversity.ummz.umich.edu
- Mammal Species of the World Web Site: http://nmnhwww.si.edu/msw/about.html
- The American Society of Mammalogists: http://www.mammalogy.org

USING THIS BOOK TO IDENTIFY A MAMMAL

1. WHAT GENERAL TYPE OF MAMMAL IS IT?
Whale? Bat? Mouse? Consult the mammal chart on the next page for hints on general mammal classification, and appropriate page numbers.

2. LOOK AT THE MAMMAL PICTURES.
Examine the artwork that covers your mammal type. Look for pictures similar to your mystery mammal.

3. CONSIDER RANGE MAPS.
Species that don't live where the mystery animal originated are unlikely, but not impossible, candidates.

4. COMPARE THE CANDIDATES.
Look carefully at the illustrations of species that resemble your mystery mammal and are known to live in your area. These should be on the same, or neighboring pages, allowing easy comparisons.

5. READ THE DETAILS.
The text for each species provides additional details that may help in species identification. If no illustrations match your mystery mammal, look here for details about known variation in size and color. If more than one species seems to match your mystery mammal, look here for details about small characters or measurements that distinguish similar species. Sometimes the distinguishing character is quite technical, and may require examining features of the bones or teeth (e.g. some shrews, some western chipmunks), which may not be practical for all situations. Don't forget to consider habitat type, which is always described in the final sentence of a paragraph, and can sometimes be quite specific and useful in identification.

6. MAKE IDENTIFICATION.
In most cases this book will help you precisely identify your mystery mammal to the species level. However, the species of some groups are very difficult to identify, requiring skeletal material, or even genetic tests (e.g. some gophers and ground squirrels). If this is the case, you may have to settle with a genus level identification (e.g. *Thomomys* sp.) or a species group identification (e.g. the Richardson's Ground Squirrel Species Group).

QUICK MAMMAL ID CHART

This chart should help point the novice mammal identifier to the right section of the book based on general characters. Based on the two descriptors, find the group or groups of mammals that best fits your mystery mammal. Then go to those pages and compare similar species.

DESCRIPTOR 1	DESCRIPTOR 2	Name	Page
FLYING OR GLIDING MAMMALS	FLY; WING MEMBRANE MOSTLY NAKED	Bats	140–156
	GLIDE; HEAVILY FURRED MEMBRANE	Flying squirrels	52
SEAGOING MAMMALS	VERY LARGE	Whales	206–212, 224–226
	SMALL TO MEDIUM-SIZED	Dolphins	214–220
	ROUNDED HIND FLIPPER	Manatee	190
	PAIRED HIND FLIPPERS	Sea otter, seals	174, 184–190
HOOFED ANIMALS	HAVE HOOVES	Ungulates	192–204

DESCRIPTOR 1	DESCRIPTOR 2	Name	Page	
CARNIVORES	DOG RELATIVES	Canids	164–168	
	CAT RELATIVES	Felids	160–162	
	SKUNKS	Skunks	180–182	
	WEASEL RELATIVES	Mustelids	174–178	
	BEARS	Ursids	170	
	RACCOON RELATIVES	Procyonids	172	
MEDIUM-SIZED MAMMALS	QUILLS	Porcupine	42	
	LARGE, FLATTENED, SCALY TAIL	Beaver	44	
	SMALL EARS, EYES, AND TAIL	Sewellel	42	

DESCRIPTOR 1	DESCRIPTOR 2	Name	Page	
MEDIUM-SIZED MAMMALS *CONT.*	AQUATIC; ROUNDED TAIL	Muskrats, Nutria	44	
	SHORT LEGS, SHORT BUSHY TAIL	Marmots	46	
	LARGE EARS	Rabbits, Hares	32–40	
	ARMORED	Armadillo	16	
	LONG, BUSHY, RINGED TAIL	Ringtail	172	
	POINTY NOSE, FAINTLY RINGED TAIL	Coati	172	
	WHITE FUR; SCALY PINKISH TAIL	Opossum	16	
SMALL MAMMALS	BUSHY TAIL	Tree squirrel, ground squirrel, woodrat	48–52, 56–66, 102–106	
	BIG-EARED AND LONG-TAILED RATS	Woodrats, *Rattus*	102–106	

DESCRIPTOR 1	DESCRIPTOR 2	Name	Page
SMALL MAMMALS CONT.	SMALL EAR, NO TAIL	Pikas	32
	HAIRY TAIL, SMALL EARS	Ground squirrels, prairie dogs	54–66
	STRIPY FACE	Chipmunks	66–74
VERY SMALL MAMMALS (MOUSE SIZE)	BIG EARS, WHITE BELLY	*Peromyscus* and relatives, *Baiomys, Reithrodontomys*	110–120
	BIG FEET, LONG TAIL	Kangaroo mice, kangaroo rats, jumping mice	90–1
	SMALL EARS AND EYES	Voles and relatives	122–138
	TINY EARS AND EYES	Shrews, moles, gophers	18–30, 76–80
	SHORT TAIL	Grasshopper mice	108
	SMALL FEET, LONG TAIL	Rice rats	108
	SMALL EARS, TAN COLOR	*Perognathus, Chaetodipus, Liomys*	82–86
	SMALLEST MICE	*Baiomys, Mus, Reithrodontomys*	110–112

HOW ARE MAMMALS RELATED?

The family tree of mammals below is arranged to show the basic phylogenetic, or evolutionary, relationships of all of the major orders of mammals. Our understanding of these relationships is changing rapidly, as mammalogists bring to bear a series of new techniques. Significant advances in molecular genetics in recent years have greatly increased our ability to study the evolution of all mammals. In the schematic diagram below, the length of the lines connecting the groups is a relative indication of how closely related each group is to another. This diagram is a composite, based on several recent studies. The actual degree of relatedness is poorly known for most groups. However, ongoing studies not only document changes, but add to our confidence in our ability to recognize phylogenetic relationships.

The study of phylogeny forms part of the field of systematic biology, and scientists doing this type of research are known as systematists, or taxonomists. Understanding systematic relationships at the species level, and at higher taxonomic levels such as the orders shown below, is fundamental to all other studies of the biology of any organism.

FAMILY TREE

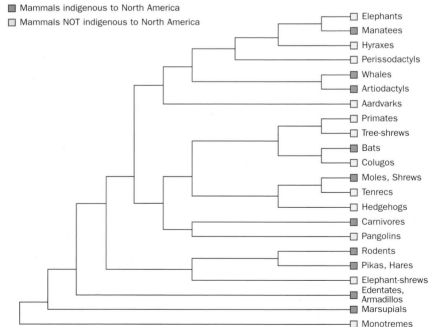

Mammal Measurements and Anatomy

Because we rely on differences in morphology to identify mammals, some knowledge of their anatomy is necessary for distinguishing similar species. Traditionally, mammalogists have used measurements of different parts of the anatomy to make these distinctions. Externally, the measurements most often used are Total Length, Tail Length, Hind Foot Length, and Ear Length. For bats, forearm length is also sometimes a useful indicator. We have also included at least Total Length, Tail Length, and Weights for each species; where useful for identification, we have included other measurements in the text. Obviously, these very specific measurements were done on mammals in the hand, and in most cases, on museum specimens. Translating them into useful gauges of the size of an animal you see scampering away from you will be much more difficult. While most measurements and descriptions we give are nontechnical, bats are so specialized that some unfamiliar terminology is needed. The figures on the right illustrate these terms.

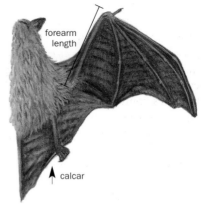

forearm
length

calcar

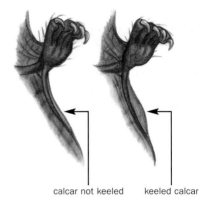

calcar not keeled   keeled calcar

── Note on measurements

The main section of the book describing each species uses metric measurements. The following conversions link metric with Imperial measurements:

10mm	=	1cm (0.4in)
100cm	=	1m (3ft 3in)
1000g	=	1kg (2.2lb)
1ha	=	2.47 acres

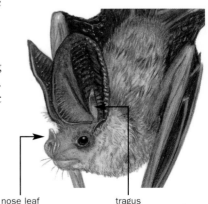

nose leaf   tragus

PLATE 1
OPOSSUM AND ARMADILLO

OPOSSUMS – Only one of the 69 species of New World marsupials extends its range northward into the United States. The Virginia Opossum is still spreading north in association with human settlements. Frostbite regularly nips off their ear and tail tips on cold winter nights.

VIRGINIA OPOSSUM *Didelphis virginiana* 350–940mm,
215–470mm, ♂ 800–6500g; ♀ 300–3700g

Unique with white head and long, scaly, prehensile tail. Medium-sized, rather ponderous-looking mammal with a long pink-tipped snout, white toes, and leathery, white-tipped ears. Body fur is gray with long white and gray guard hairs overlying underfur, frequently yielding overall scruffy appearance. Known for "playing possum," a catatonic state assumed in the face of danger. Although it is omnivorous, its slow reflexes make it a better scavenger than active hunter. Nocturnal, both terrestrial and arboreal, the opossum favors woodland habitats, but is a successful invader of urban and suburban areas as well.

ARMADILLOS – Only one of the 20 New World armadillos inhabits the United States. This insectivorous group specializes on ants and termites, and all species have very reduced, peglike dentition.

NINE-BANDED ARMADILLO *Dasypus novemcinctus* 615–800mm,
245–370mm, ♂ 5.5–7.7kg; ♀ 3.6–6.0kg

Bony skin plates are unique. Long head, prominent ears, short legs, and short, tapered, scaly plated tail are distinctive. Body has nine moveable bands encircling the mid-section. It digs burrows with its nose and forefeet, and lines them with vegetation. Primarily nocturnal and crepuscular, it is sometimes active during the day. All litters consist of a set of identical quadruplets that come from a single egg. Although clumsy looking when rooting through leaf litter for insects, it can escape quickly by bounding straight up and then running away quickly. It is adaptable to a wide variety of habitats, and limited only by colder weather in the north.

juveniles

hanging from
prehensile tail

VIRGINIA OPOSSUM

NINE-BANDED ARMADILLO

PLATE 2
EASTERN *SOREX*

SOREX SHREWS – This group of long-tailed shrews are all terrestrial insect eaters. Because of their small size, most are rarely captured in box traps. Pitfall traps have revealed surprising diversity and abundance in many areas. Many of these species are difficult to identify, even for experts. Geographic range, color, and tail distinguish most species, but dental characters are needed for some.

CINEREUS SHREW *Sorex cinereus* 75–125mm, 30–50mm, 2–5g

Tip of tail black. Medium-sized shrew lacking distinctive markings. Back brownish, fur of underparts is gray at base, paler at the tips. Darker color in winter. Tail is long (roughly 40% of total length), brown above and slightly paler below; dark tail tip not always obvious. (For dental characters see page 22.) Nocturnal and rarely seen. Common in northern forests, shrublands, grassy areas, and herbaceous habitats.

SOUTHEASTERN SHREW *Sorex longirostris* 75–100mm, 25–40mm, 2–6g

Only long-tailed shrew in most of range. Small brown body with long and narrow rostrum. Broadly distributed in Southeastern United States, but rarely seen. Favors moister areas with dense ground cover, but can occur in pine woods and scrub habitats as well.

PYGMY SHREW *Sorex hoyi* 60–105mm, 20–40mm, 2–7g

Our smallest mammal, with small, bright eyes and obscure ear pinnae. Long snout has conspicuous whiskers. Color varies from coppery brown (summer) to grayish (winter) above, underparts are paler, grayish brown or drab tinged with copper or tan. Tail is dark brown above, paler below, and muzzle is paler than the crown and back. Tail is less than 40% of total length. (For dental characters see page 22.) Often stands on hind limbs in kangaroo fashion, and runs quickly with the extended tail slightly curved. Most abundant in boreal habitats.

LONG-TAILED SHREW *Sorex dispar* 105–135mm, 45–65mm, 3–8g

Slender, dark-gray shrew with a long tail. Belly only slightly paler than back, tail not distinctly bicolored. Smaller than Water Shrew; larger than Gaspé Shrew; tail longer than Smoky Shrew. Typically smaller in northern parts of range. Rarely encountered due to activity in passages and runways below rocky surface. Procumbent incisors may aid in probing rocky crevices for insects. Favors rocky areas in cool, moist, boreal forests.

SMOKY SHREW *Sorex fumeus* 110–125mm, 35–50mm, 6–11g

Short, heavy-bodied shrew with a dark back, light belly, and long tail. Back is gray in winter and brown in summer. Tail shorter than Water Shrew and Long-tailed Shrew. Nocturnal and active year-round. Nests of shredded vegetation are often in rotten logs or under rocks. Common in wet forests, where it is active on forest floor and under litter.

GASPÉ SHREW *Sorex gaspensis* 95–125mm, 45–55mm, 2–4g

Very limited range in Quebec and Nova Scotia. Like Long-tailed Shrew but smaller, with paler pelage. Almost never seen, except in pitfall traps. Frequents rocky boreal habitats preferring cool, rocky stream edges.

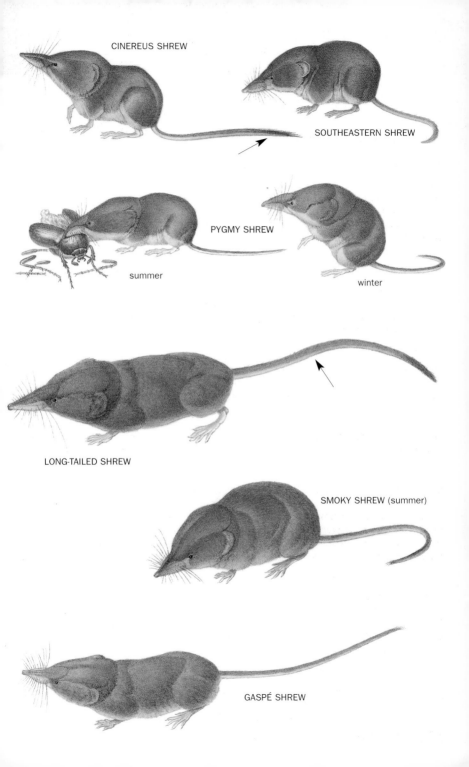

CINEREUS SHREW

SOUTHEASTERN SHREW

PYGMY SHREW

summer

winter

LONG-TAILED SHREW

SMOKY SHREW (summer)

GASPÉ SHREW

PLATE 3
OBVIOUS WESTERN SOREX

WATER SHREW *Sorex palustris* 130–170mm, 57–89mm, 8–18g

Large shrew with black or grayish back. The long (more than 18mm) hind feet have fringes of stiff hairs. Almost always found in or near streams or lakes. Capable of skipping across surface of water and diving to bottom in search of food. Common in boreal forests.

MARSH SHREW *Sorex bendirii* 128–174mm, 58–80mm, 7–21g

Large, velvety, brownish-grayish dark shrew with relatively uniform belly and back. Similar to Water Shrew, but without stiff hairs on hind toes. Forages both on land and in water, but eats prey only on land. Occupies coastal forests from British Columbia to San Francisco.

ARCTIC SHREW *Sorex arcticus* 100–125mm, 35–45mm, 5–13g

Dark tricolored pelage. Back is very dark brown to black, sides are paler brown, and belly is grayish brown. Winter pelage is more uniformly dark brown, and young of the year may be bicolored (brownish above with pale brown underparts). The tail is indistinctly bicolored, brown to brownish black above and paler below. The population from New Brunswick and Nova Scotia has recently been described as a distinct species (*Sorex maritimensis*). Active round the clock, but more so at night and less so 6–10 A.M. Occurs mainly in marshes and grassy clearings in northern boreal coniferous forests.

TUNDRA SHREW *Sorex tundrensis* 85–120mm, 20–35mm, 5–10g

Light tricolored pelage. Light brown back contrasts with pale brown or grayish sides and pale belly. Larger than Barren Ground Shrew, lighter colored than Arctic Shrew with a relatively shorter tail than both. Winter pelage is longer and more uniform in color, although underparts are always paler than back. Females produce several litters between May and September. Occurs in dense vegetation including grasses, shrubs, and dwarf trees in tundra zone north of boreal forests.

BARREN GROUND SHREW *Sorex ugyunak* 75–105mm, 20–30mm, 3–5g

Small tricolored shrew. Distinct line separates dark fur on back from pale fur on sides. Once thought to be a subspecies of *S. cinereus*, but the coloration is more like *S. tundrensis*, which is much larger. The brown back of the Barren Ground Shrew looks like a well-defined median dorsal stripe. Tail is distinctly bicolored, pale brown above and whitish below, with a pale buff to brown terminal tip. Immatures are much more uniformly colored than adults. Lives in short grass meadows and woody thickets of northern tundra.

PRIBILOF ISLAND SHREW *Sorex pribilofensis* 90–95mm, 30–35mm, 3–4g

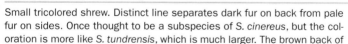

Only shrew on St. Paul Island. Superficially similar to the closely related Barren Ground Shrew. These small, bicolored shrews have been isolated in the Pribilof Islands for about 16,000 years. It may have occurred at one time on Unalaska Island also, but no recent specimens are known. Nothing is known about the life history of this species.

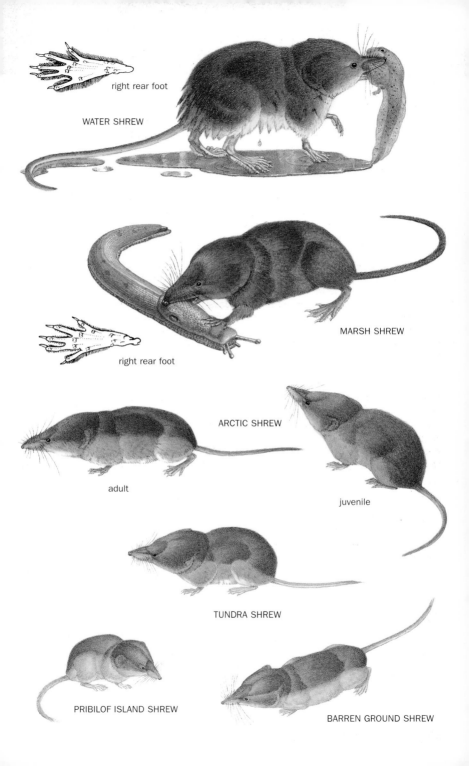

right rear foot

WATER SHREW

right rear foot

MARSH SHREW

ARCTIC SHREW

adult

juvenile

TUNDRA SHREW

PRIBILOF ISLAND SHREW

BARREN GROUND SHREW

PLATE 4
INLAND *SOREX* WITH SKULLS

CINEREUS SHREW *Sorex cinereus*

Skull has a series of upper unicuspid teeth showing a gradual reduction in size from front to back. (*See* page 18 for more details.)

PYGMY SHREW *Sorex hoyi*

A tiny shrew with upper unicuspate teeth including a small peglike posterior unicuspid, and a third unicuspid that is a tiny disk, barely visible to the naked eye. (See page 18 for more details.)

PRAIRIE SHREW *Sorex haydeni* 75–100mm, 25–40mm, 2–5g

Small brown shrew with relatively short tail lacking black tip. Red pigmentation on lower incisor on tip and first two cusps, with more on only the third cusp. Occupies prairie grasslands in the Great Plains, including the Black Hills. Builds bird-like nests under logs and rocks.

DWARF SHREW *Sorex nanus* 80–105mm, 25–45mm, 2–3g

Tiny shrew with pelage back black at base, overlain with grayish brown. Third upper unicuspid smaller than second and fourth; fifth minute. Occurs from arid shortgrass prairies to alpine tundra throughout Rocky Mountain area. Prefers rocky areas such as talus slopes.

MERRIAM'S SHREW *Sorex merriami* 90–105mm, 35–40mm, 4–7g

Small shrew with grayish-brown pelage on back, paler flanks, and whitish underparts. Third upper unicuspid larger than fourth; fifth tiny. No medial tines on incisors. Occurs in sagebrush, grasslands, and woodlands up to 3000m. May use vole runways in grassy areas.

ARIZONA SHREW *Sorex arizonae* 80–115mm, 35–55mm, 2–5g

Slightly larger than Merriam's Shrew and slightly smaller than Montane Shrew, with similar coloring. Third unicuspid roughly equal in size to fourth and a small medial tine on upper incisors. Favors forested slopes from 1500–2500m. Active year-round, but more so during periods of rainfall.

PREBLE'S SHREW *Sorex preblei* 75–95mm, 30–40mm, 2–4g

Very small with gray-brown back and silvery belly. Distinguished by medial tine on upper incisors, and the third unicuspid being equal in size to fourth. Among the smallest shrews in North America, it occupies shrub and grasslands, as well as wetter areas at intermediate elevations.

MT. LYELL SHREW *Sorex lyelli* 88–108mm, 38–43mm, 4–5g

Similar to Cinereus Shrew, but restricted to central Sierra Nevada Mountains of California (red area on map). Third unicuspid larger than fourth.

INYO SHREW *Sorex tenellus* 85–105mm, 35–50mm, 3–4g

Slightly larger than Dwarf Shrew, and with relatively longer tail. Grayish-brown pelage above and paler below; sometimes has reddish color on back. Rarely trapped, almost nothing is known about habits. Occurs in semiarid areas and some moist woodlands on mountain ranges in the Great Basin (see yellow area on map of Mt. Lyell Shrew).

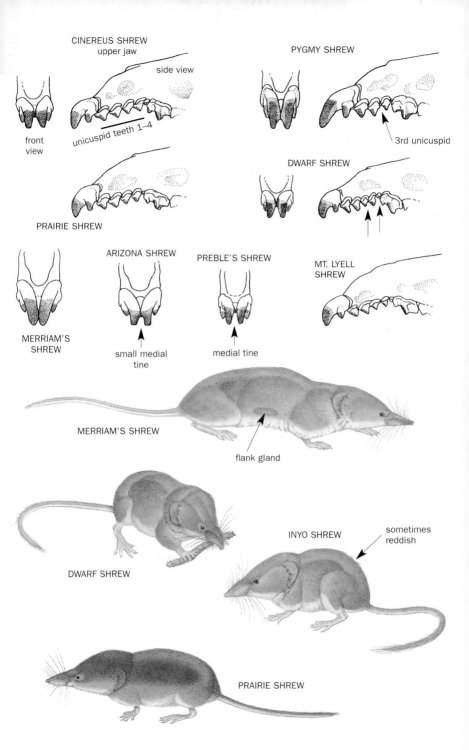

CINEREUS SHREW
upper jaw

side view

front
view

unicuspid teeth 1–4

PYGMY SHREW

3rd unicuspid

PRAIRIE SHREW

DWARF SHREW

ARIZONA SHREW

PREBLE'S SHREW

MT. LYELL
SHREW

MERRIAM'S
SHREW

small medial
tine

medial tine

MERRIAM'S SHREW

flank gland

DWARF SHREW

INYO SHREW

sometimes
reddish

PRAIRIE SHREW

PLATE 5
COASTAL *SOREX* WITH SKULLS

MONTANE SHREW *Sorex monticolus* 95–140mm, 30–60mm, 4–10g

Dark brown pelage; medial tine on upper incisors large. Third upper unicuspid smaller than fourth. Red pigmentation extends above notch on upper incisors. Hind feet have more (more than four) paired friction pads than Vagrant Shrew. Active all year at all hours. Widespread in Northwest North America and Rocky Mountains, but restricted to wet habitats.

ORNATE SHREW *Sorex ornatus* 80–110mm, 28–46mm, 3–9g

Small, grayish-brown shrew with medial tine on upper incisors. Fourth unicuspid larger than third. The belly is paler than the back and the skull lacks post-mandibular foramina. Body is like Inyo Shrew, but slightly larger and darker. Restricted to Pacific Coastal region of southern California. Favors streamsides with dense vegetation, but also occurs in upland woodlands and forests.

BAIRD'S SHREW *Sorex bairdii* 100–145mm, 30–65mm, 5–11g

Externally like Vagrant Shrew but slightly larger, with a well pigmented medial tine on the upper incisors. Third upper unicuspid larger than fourth. Smaller than Marsh Shrew and Fog Shrew; larger than Montane Shrew. Pelage darker than Pacific Shrew. In winter the pelage is darker brown. Favors moist coniferous forests (see red area on map of Fog Shrew).

FOG SHREW *Sorex sonomae* 105–180mm, 36–85mm, 5–15g

Largest of the brown shrews found along the Pacific coast. Back color is dark grayish brown, and the upper incisors lack medial tines. Like Pacific shrew but larger, browner, and with no protuberances on upper incisors. Restricted to fog zone of Oregon and California coasts (yellow area on map). Forages at night by searching through the leaf litter and surface soil for invertebrates.

VAGRANT SHREW *Sorex vagrans* 100–115mm, 38–48mm, 3–8g

Has small, pigmented medial tines on upper incisors that are usually separated from the pigmented tips by a pale line. Back is brown, sides are paler, and belly is white; tail is bicolored. Prefers moist habitats, frequently in sedges, grasses, and willows along streams and lakes, and in coastal salt marshes.

PACIFIC SHREW *Sorex pacificus* 135–155mm, 60–70mm, 10–18g

Large, pale, reddish-brown shrew. Hind feet with five sets of friction pads. Upper incisors do not have a medial tine, but do have a small protuberance. Nests are constructed from vegetation, and latrines are maintained away from the nest. Prefers moist areas along streams, especially in dense vegetation with fallen logs.

TROWBRIDGE'S SHREW *Sorex trowbridgii* 104–131mm, 48–59mm, 3–5g

Small shrew with nearly uniform, dark fur, whitish feet, and bicolored tail. This is the only shrew in its range whose belly is not significantly paler than its back, but does have a strongly bicolored tail. Grayer in winter and browner in summer. Preferred habitat includes litter with thick organic layer, low water table, and dense ground cover or canopy.

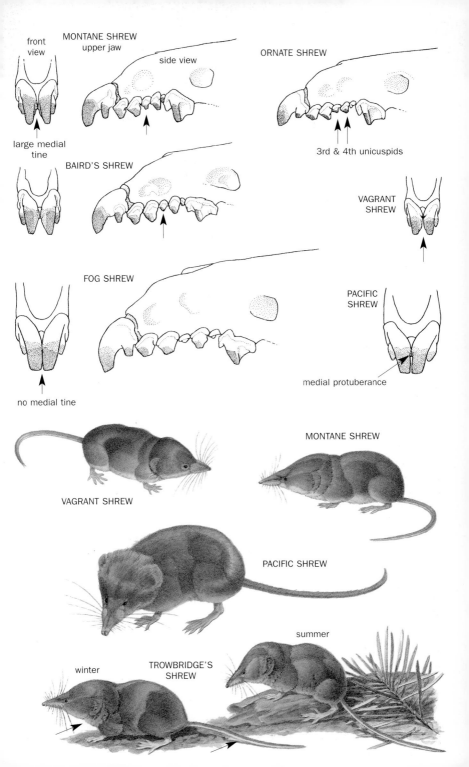

front view

MONTANE SHREW
upper jaw

side view

large medial tine

ORNATE SHREW

3rd & 4th unicuspids

BAIRD'S SHREW

VAGRANT SHREW

PACIFIC SHREW

FOG SHREW

no medial tine

medial protuberance

VAGRANT SHREW

MONTANE SHREW

PACIFIC SHREW

winter

TROWBRIDGE'S SHREW

summer

PLATE 6
OTHER SHREWS

LEAST SHREW *Cryptotis parva* 61–89mm, 19–37mm, 3–10g

Tiny shrew with a very short tail (less than 45% of head and body length). Smaller and browner than other short-tailed shrews in their range. Nests are constructed of grass and leaves in hidden areas, and may contain up to 31 individuals. Occurs in wide variety of habitats, including grassy, weedy, and brushy fields, marshes, and wooded habitats.

SHORT-TAILED SHREWS *BLARINA* – All short-tailed shrews have nearly uniform silver to black fur with brown tips on the hairs. Summer fur is shorter and slightly paler. They are significantly larger and have shorter tails than other shrews in their range. The three species of short-tailed shrews are very similar, and most easily identified by geography. In areas of overlap, the Northern is usually the largest, and the Southern the smallest. Eliot's is intermediate in size, but there is considerable variation and overlap. The three are identified confidently only by examining chromosome numbers. They are habitat generalists, but are most abundant in moist, well-drained areas.

NORTHERN SHORT-TAILED SHREW *Blarina brevicauda*
95–139mm, 17–32mm, 11–30g

SOUTHERN SHORT-TAILED SHREW *Blarina carolinensis*
72–107mm, 12–26mm, 5–13g

(For distribution see yellow area on map.)

ELIOT'S SHORT-TAILED SHREW *Blarina hylophaga*
92–121mm, 19–25mm, 13–16g

(For distribution see red area on map of Southern Short-tailed Shrew.)

DESERT SHREW *Notiosorex crawfordi* 77–98mm, 24–32mm, 3–6g

Conspicuous ears extend beyond the silvery to brownish-gray fur. Tail short (less than one third of total length) and unicolored. The belly is paler. This is the only shrew in most of the arid and semiarid habitats of the southwestern United States. It builds golf-ball-sized nests of fine fibers inside woodrat houses. Ranges widely in arid environments from southern California eastward to Arkansas.

LEAST SHREW

winter

summer

summer

SHORT-TAILED SHREWS

winter

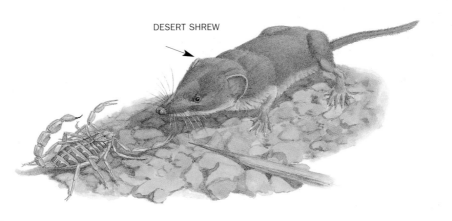

DESERT SHREW

PLATE 7
WESTERN MOLES

MOLES – Moles are tunneling insect- and worm-eaters with tiny eyes. They differ from shrews in having broad forepaws, lacking pigmentation on their teeth, and generally being larger. Compared with the urinary papilla of a female, the penis of a male mole is larger and starts further from the anus. Evidence of their digging in an area can often be seen in the form of tunnels near the surface, and piles of excavated dirt mounded up above tunnel entrances.

AMERICAN SHREW MOLE *Neurotrichus gibbsii* 92–132mm, 12–19mm, 9–11g

The Shrew Mole is the smallest of the moles, with a short thick tail. Forepaws larger than in any shrews. Some of the hairs are longer and coarser than others, making them less velvety looking than other moles. It spends more time above ground than other moles, and frequently makes tunnels through the leaf litter rather than completely underground. Prefers areas with heavy leaf litter, or abundant shrub or bunchgrass cover.

TOWNSEND'S MOLE *Scapanus townsendii* ♂ 207–237mm, 35–56mm, 100–171g; ♀ 183–209mm, 29–51mm, 50–110g

Larger than all other moles in North America. Hindfoot exceeds 24mm; tail not as hairy as the Broad-footed Mole. May form as many as 800 large mounds per hectare in some areas. Exceptionally large mounds contain a large nest chamber 15–20cm below the surface. Favors deep, loamy soils in meadows and adjacent areas.

BROAD-FOOTED MOLE *Scapanus latimanus* 136–193mm, 21–45mm, 39–55g

Hairiest tail of western moles. The snout is shorter and broader than other moles, and the unicuspid teeth are unevenly spaced. Dark gray fur may have coppery wash. Shallow foraging tunnels are constructed near the surface, and deeper, more permanent tunnels are used for resting and nest building. Prefers moist soils from sea level up to 3000m.

COAST MOLE *Scapanus orarius* ♂ 136–190mm, 30–45mm, 64–91g; ♀ 133–168mm, 21–46mm, 61–79g

The forefoot is broader than it is long, and the hindfoot is less than 24mm long. Evenly spaced unicuspids and scantily haired tail distinguish it from Broad-footed Mole, and it is slightly smaller than Townsend's Mole. Mounds of soil deposited on the surface from tunneling activities are slightly smaller than those of Townsend's Moles. Occurs in coastal sand dunes, grassy meadows, sagebrush grasslands, deciduous forest, and coniferous woodlands.

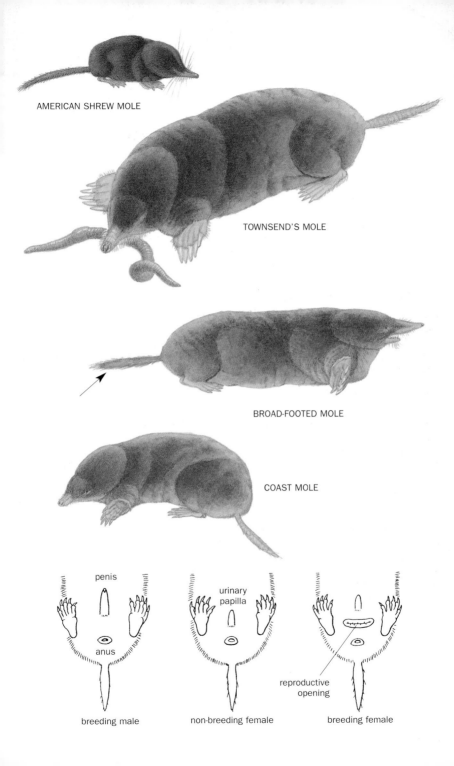

AMERICAN SHREW MOLE

TOWNSEND'S MOLE

BROAD-FOOTED MOLE

COAST MOLE

penis

anus

breeding male

urinary
papilla

non-breeding female

reproductive
opening

breeding female

PLATE 8
EASTERN MOLES

EASTERN MOLE *Scalopus aquaticus* ♂ 103–208mm, 16–38mm, 40–140g; ♀ 129–168mm, 20–28mm, 32–90g

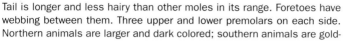

Tail is longer and less hairy than other moles in its range. Foretoes have webbing between them. Three upper and lower premolars on each side. Northern animals are larger and dark colored; southern animals are golden or silvery colored. Active year-round, and feed on a variety of invertebrates including earthworms and ant larvae. Both surface and deeper tunnels are constructed in moist, loamy soils throughout eastern North America.

HAIRY-TAILED MOLE *Parascalops breweri* 151–173mm, 26–33mm, 41–63g

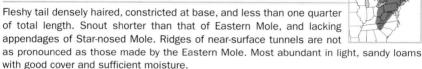

Fleshy tail densely haired, constricted at base, and less than one quarter of total length. Snout shorter than that of Eastern Mole, and lacking appendages of Star-nosed Mole. Ridges of near-surface tunnels are not as pronounced as those made by the Eastern Mole. Most abundant in light, sandy loams with good cover and sufficient moisture.

STAR-NOSED MOLE *Condylura cristata* 132–230mm, 48–99mm, 40–85g

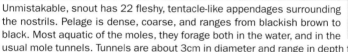

Unmistakable, snout has 22 fleshy, tentacle-like appendages surrounding the nostrils. Pelage is dense, coarse, and ranges from blackish brown to black. Most aquatic of the moles, they forage both in the water, and in the usual mole tunnels. Tunnels are about 3cm in diameter and range in depth from 3 to 60cm. Tails thicken in winter for fat storage. May be gregarious or even possibly colonial. Prefers wet areas.

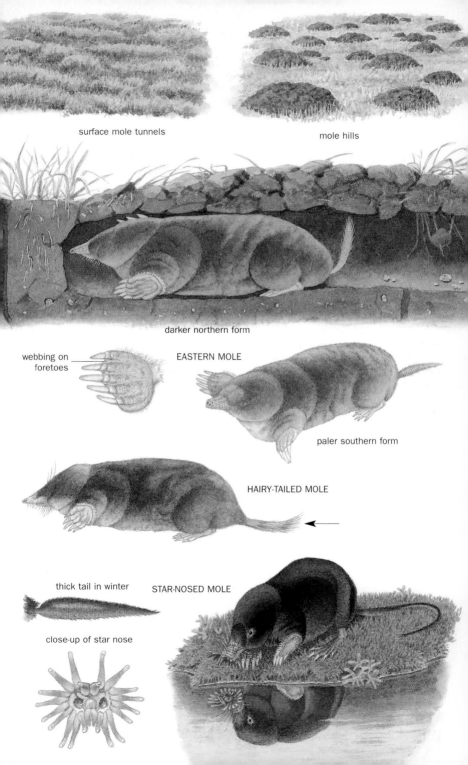

surface mole tunnels

mole hills

darker northern form

EASTERN MOLE

webbing on foretoes

paler southern form

HAIRY-TAILED MOLE

thick tail in winter

STAR-NOSED MOLE

close-up of star nose

PLATE 9
PIKAS AND WESTERN COTTONTAILS

PIKAS – These egg-shaped rabbit relatives have small ears and no apparent tail. Our two American species do not overlap in range. Signs includes hay piles in crevices and urine stains on rocks. Emit loud, short, sharp calls for alarm and social purposes. Use talus slopes or areas of broken rocks near meadows.

COLLARED PIKA *Ochotona collaris* 178–198mm, 0mm, 117–145g

Has gray collar, creamy-buff facial patch, and white underparts. Upperparts are drab and washed with gray or black. July juveniles are near adult-size, but are clear gray while adults' heads and necks are tinged with brown at this time. Winter pelage is much longer than summer pelage.

AMERICAN PIKA *Ochotona princeps* 162–216mm, 0mm, 121–176g

Lacks gray collar, has rusty brown facial patch, and buff underparts. Summer pelage ranges from grayish to cinnamon-brown. Winter pelage is grayer and nearly twice as long. Soft gray juveniles mature to adult color by late summer. A long call is emitted by males in mating season.

COTTONTAILS AND PYGMY RABBIT – Typical rabbits, often identified by relative ear size. Cottontail identification in some parts of the Northeast may require examination of skull characters.

MOUNTAIN COTTONTAIL *Sylvilagus nuttallii* 338–390mm, 30–54mm, 628–687g

Short, rounded ears are haired on inner surface; whiskers are mostly white. Hind legs are long; feet covered with long, dense hair. Large, grizzled tail is dark above and white below. Associated with sagebrush or timbered areas.

DESERT COTTONTAIL *Sylvilagus audubonii* 372–397mm, 45–60mm, 755–1250g

Large ears are pointed and sparsely haired. Slender feet lack dense hair of other *Sylvilagus*. Large legs. Whiskers are generally black. Large tail is dark above and white below. Crepuscular. Occurs in desert brush, near willows along rivers, or in pinyon-juniper woodlands with shrub cover.

EASTERN COTTONTAIL *Sylvilagus floridanus*

Larger than other western cottontails, relatively smaller ears than Desert Cottontail. (*See* page 36 for more details.)

COLLARED PIKA

winter

AMERICAN PIKA

summer

MOUNTAIN
COTTONTAIL

DESERT COTTONTAIL

EASTERN COTTONTAIL (western form)

PLATE 10
PYGMY RABBITS AND SHORT-EARED COTTONTAILS

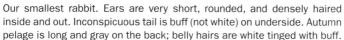

PYGMY RABBIT *Brachylagus idahoensis* ♂ 252–285mm, 15–20mm, 373–435g; ♀ 320–305mm, 15–24mm, 415–458g

Our smallest rabbit. Ears are very short, rounded, and densely haired inside and out. Inconspicuous tail is buff (not white) on underside. Autumn pelage is long and gray on the back; belly hairs are white tinged with buff. Mid-winter fur is worn and gray, becoming darker gray in spring and summer. Moves by scurrying, rather than hopping. Mostly crepuscular. Alarm call is a buzzing, one- to seven-syllable squeal. Specialist of big sagebrush habitat.

BRUSH RABBIT *Sylvilagus bachmani* 303–369mm, 10–30mm, 511–917g

Small rabbit with short legs and tail. Ears are slightly pointed and sparsely haired inside (not well haired like Pygmy Rabbit). Dark gray on back and sides; pale gray on belly and underside of tail. Whiskers are mostly black, some may have white tips. Will forage in groups. Thumps ground with hind foot when frightened; may climb low branches to escape. Californian subspecies (*S. b. riparius*) is Endangered. Lives in dense brush.

SWAMP RABBIT *Sylvilagus aquaticus* 452–552mm, 50–74mm, 1.6–2.7kg

Our largest rabbit; ears are relatively small. Underside of tail is white. Head, back, and upper tail are rusty brown to black; throat and belly are white. Cinnamon eye rings. Feet are pale to reddish. Most active at dusk. Good swimmer. Common in swampy areas.

MARSH RABBIT *Sylvilagus palustris* 425–440mm, 33–39mm, 1.2–2.0kg

Smaller than Swamp Rabbit, with a dingy underside of tail (rarely white). Dainty feet are red to buff in color. Back, rump, upper tail, and hind legs are chestnut brown to rusty red; back of neck is dark cinnamon; abdomen is white, rest of belly is buff to brown. Most active at night. Florida Keys subspecies (*S. p. hefneri*) Endangered. Uses swamps, lake borders, and other wet areas.

PYGMY RABBIT

BRUSH RABBIT

SWAMP RABBIT

MARSH RABBIT

PLATE 11
EASTERN COTTONTAILS

EASTERN COTTONTAIL *Sylvilagus floridanus* 40–50cm, 2–6cm, 800–1500g

Our most common cottontail; has relatively large ears. Different from Appalachian and New England Cottontails by: 1) often showing a white (never black) spot between ears; 2) lacking a black fringe on front edge of ears; 3) lacking a black penciled effect on back; 4) having a postorbital process that is broad, flat, and frequently touching the skull (not thin, tapering, and rarely touching skull); 5) having a suture between nasals and frontals that is smooth (not jagged or irregular). Upperparts are densely furred in brownish or grayish; belly and undertail are furred in white. Common in variety of habitats including overgrown fields, meadows, and brushy areas. (*See* also page 32 for western form.)

NEW ENGLAND COTTONTAIL *Sylvilagus transitionalis* 40–44cm, 5–7cm, 995–1347g

Restricted to New England. Virtually identical to the Appalachian Cottontail, distinguished only by range and genetics. Identification vs. Eastern Cottontail outlined above. Nests are in depressions, ca. 12cm deep by 10cm wide and lined with grass and fur. Eats grasses and clovers in summer and forbs and twigs in winter. Prefers forested habitats and rarely ventures into the open.

APPALACHIAN COTTONTAIL *Sylvilagus obscurus* 39–43cm, 2–6cm, 756–1038g

Restricted to Appalachian highlands. Identification vs. Eastern Cottontail outlined above. Differentiated from New England Cottontail only by range and genetics. Upperparts are pinkish buff; belly is white or buff. Back is overlaid with a black wash, creating a penciled look. Cheeks are grizzled. Rounded ears are fringed with black hairs. Usually shows a distinct black (never white) spot between ears. Occurs with dense cover in high elevation boreal forests.

CAPE HARE *Lepus capensis* 64–70cm, 7–10cm, 3.0–5.6kg

Non-native species with declining distribution. Different from native *Lepus* in having dorsal hairs white at base. Rusty- to yellow-brown upperparts in summer with black hairs interspersed; white below. Grayish upperparts in winter. Center of tail black above; tail all white below. Predominantly nocturnal. Populations are presently known from Ontario and the Hudson River Valley. Uses open, cultivated fields.

EUROPEAN RABBIT *Oryctolagus cuniculus* 34–45cm, 4–8cm, 1.3–2.2kg

A small, non-native species restricted to islands of Pacific coast. Upperparts generally grayish and interspersed with black, brown, and sometimes red hairs. Underparts are pale gray. Tail almost completely white, with some dark hairs on upperside. Relatively short ears lack black tips. Domestic breeds vary from black to white. Mainly crepuscular.

post-orbital process

nasal/frontal suture

New England Cottontail

Eastern Cottontail

top of skull

EASTERN
COTTONTAIL

NEW ENGLAND
COTTONTAIL

APPALACHIAN
COTTONTAIL

Eastern
Cottontail

New
England
Cottontail

CAPE HARE

EUROPEAN
RABBIT

winter

summer

PLATE 12
NORTHERN HARES

NORTHERN HARES – These snow-adapted species are generally brown in summer and white in winter. Their smaller ears distinguish them from jackrabbits. Skull characters are needed to distinguish some species in winter.

SNOWSHOE HARE *Lepus americanus* ♂ 40–44cm, 2–5cm, 0.9–1.7kg; ♀ 42–52cm, 3–8cm, 0.9–2.2kg

Smaller than other *Lepus*; base of winter hairs brown (not white). Summer pelage rusty brown above and grayish below; ears tipped in black; nostrils and tail white. Winter pelage white with black-tipped ears and yellowish underpaws. Some populations in Oregon and Washington are brown all year. Mostly nocturnal. Uses dense thickets in coniferous and mixed forests in the north, and deciduous forests in the south of its range.

ALASKAN HARE *Lepus othus* 56–59cm, 6–10cm, 3.9–7.2kg

Restricted to coastal Alaska. Larger than Snowshoe Hare with longer ears (more than 73mm) and winter pelage that is completely white to the base, except black ear tips. Summer pelage is reddish brown or brownish gray with a white or gray tail. Distinguished from Arctic Hare by geographic range, by brownish (not grayish or white) summer pelage, and by having more strongly recurved incisors, a heavier rostrum, and a longer upper tooth row. Inhabits tundra and dense alder thickets.

ARCTIC HARE *Lepus arcticus* 56–63cm, 4–10cm, 2.5–3.8kg

Restricted to northeastern Canada. Larger than Snowshoe Hare with winter pelage that is completely white to the base, except black ear tips. Summer pelage gray in southern subspecies; remains white in northern subspecies. Distinguished from Alaskan Hare by range, summer color, and skull characters described above. Lives in tundra, may retreat below timberline in winter.

WHITE-TAILED JACKRABBIT *Lepus townsendii* ♂ 56–62cm, 7–10cm, 2.6–4.3kg; ♀ 58–65cm, 7–10cm, 2.5–4.3kg

White phase White-tailed Jackrabbits are distinguished from Snowshoe Hares by larger body size, longer ears, and hair that is dark at the base. White winter pelage may be tinged with buff on ears, face, back, and feet. Ear tips are black. Tail is white, sometimes with a buff dorsal stripe. Southern populations remain brown in winter. (See page 41 for summer pelage.)

SNOWSHOE HARE

winter

summer

ALASKAN HARE

summer

ARCTIC HARE

summer

winter

Arctic Hare

more recurved incisors

Alaskan Hare

winter

WHITE-TAILED JACKRABBIT

PLATE 13
JACKRABBITS

JACKRABBITS – The enormous ears of these hares help them keep cool in their hot, arid habitats.

ANTELOPE JACKRABBIT *Lepus alleni* 55–67cm, 5–8cm, 2.7–5.9kg

Enormous ears tipped in white. Only found in Mexico and southern Arizona. Back is yellowish brown darkened with black hairs; sharply demarcated sides are gray; underparts are white. Throat patch is orangish yellow. Large ears nearly naked except long white hair fringing the edge. Nocturnal and crepuscular. Does not drink water. Favors desert plains with grasses, mesquites, and *Acacia* but can persist in areas with little vegetation.

BLACK-TAILED JACKRABBIT *Lepus californicus* 46–63cm, 5–11cm, 1.3–3.3kg

Black upper tail and ear tips. Upperparts and sides are brown to dark gray, underparts pale gray. Tail gray below and black above; black color may extend as a line up lower back. Mostly nocturnal. Will feed in groups. Introduced to many eastern states. Uses agricultural and range lands, especially areas with cacti and low shrubs.

WHITE-TAILED JACKRABBIT *Lepus townsendii* ♂ 56–62cm, 7–10cm, 2.6–4.3kg; ♀ 58–65cm, 7–10cm, 2.5–4.3kg

Tail is white, sometimes showing a buff dorsal stripe. Ear tips are black. Summer upperparts are yellowish brown (*campanius* subspecies east of continental divide) or grayish brown (*townsendii* subspecies west of divide); underparts are white or pale gray with a darker throat. Northern populations molt to white in winter (see previous plate). Nocturnal. Introduced to Wisconsin. Uses open grassland, sagebrush, and meadows, especially on mountain slopes and ridges.

WHITE-SIDED JACKRABBIT *Lepus callotis* ♂ 52–53cm, 5–9cm, 1.5–2.2kg; ♀ 54–57cm, 5–9cm, 2.5–3.2kg

Only found in Mexico and Hidalgo County, New Mexico. White sides unique. Back is pale brownish red; sides, rump, thighs, and underparts are white. Black hairs are mixed throughout upperparts. Tail is white below and black above; some black tail hairs are tipped in white. White-tipped ears are sparsely haired with a dusky spot along posterior border. Nocturnal. Prefers level desert grasslands with little shrub cover.

ANTELOPE JACKRABBIT

BLACK-TAILED JACKRABBIT

WHITE-SIDED JACKRABBIT

WHITE-TAILED JACKRABBIT
summer

PLATE 14
PORCUPINE AND *APLODONTIA*

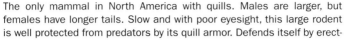

NORTH AMERICAN PORCUPINE *Erethizon dorsatum* 60–130cm, 17–25cm, 3.5–18kg

The only mammal in North America with quills. Males are larger, but females have longer tails. Slow and with poor eyesight, this large rodent is well protected from predators by its quill armor. Defends itself by erecting (not throwing) its quills, lowering its head, and backing up toward the intruder with its tail flailing. Dens in burrows, rocky crevices, and hollow trees. Eats a variety of plant material, but is especially fond of the cambium layer of coniferous trees. Will strip bark of trees and leave piles of feces beneath. Extirpated from many areas in the eastern and midwestern United States. Recently reintroduced in some areas. Rare to common in a variety of habitat types including forest, tundra, chaparral, and rangelands.

SEWELLEL *Aplodontia rufa* 24–47cm, 19–55mm, 0.8–1.2kg

An odd, medium-sized burrowing rodent with small eyes and long whiskers. The fur is dark brown, and there is a pale spot below each ear. Somewhat resembles a giant pocket gopher. The tail is short and furred. Although it rarely ventures far from its 15cm burrow entrance, it can climb trees in search of food. Strictly vegetarian, it is known for eating plants such as rhododendron and stinging nettle that other animals typically avoid. Extensive burrow systems include a toilet chamber. Typically remains underground in winter, eating cached foods. The sole member of the rodent family Aplodontiidae. Endangered in California. Poorly named, this primitive rodent is neither aquatic nor fond of mountains; it spends most of its life in underground borrows dug into the soil of moist forests.

NORTH AMERICAN PORCUPINE

SEWELLEL

PLATE 15
LARGE AQUATIC RODENTS

ROUND-TAILED MUSKRAT *Neofiber alleni* 16–23cm, 10–17cm, 200–300g

A small muskrat with a round black tail. Pelage is glossy rich brown to black, with dense underfur that is gray to brown on the back, grading to grayish or buff on the belly. Smaller than the Common Muskrat, but much larger than any mice or voles. Builds dome-shaped grass houses 18–60cm in diameter at the surface of the water. Nocturnal feeders on aquatic grasses. Lives in freshwater marshes in Florida and Georgia.

MUSKRAT *Ondatra zibethicus* 41–62cm, 18–30cm, 700–1800g

A medium-sized, brown, aquatic rodent. Back is dark brown, underside is slightly paler. Black tail is vertically flattened. Long, coarse, glossy guard hairs cover the short, dense, silky underfur. Partially webbed hind feet are larger than the forefeet. Fringes of stiff hairs along the sides of the toes further enhance swimming ability. Larger-bodied in the north of their range. Clearly larger than the Round-tailed Muskrat and smaller than the Nutria or Beaver. Uses cut vegetation to build rounded houses about 2m in diameter and 1m high; also dens in holes dug into the shore. Crepuscular and nocturnal feeder on a variety of aquatic plants. Common in brackish and freshwater lakes, rivers, and swamps.

NUTRIA *Myocastor coypus* 86–106cm, 30–43cm, 6.7–9.0kg

A large, brown, aquatic rodent with a rounded tail. Larger than muskrats and smaller than Beaver. Typically burrows into banks, but also eats and rests on small platforms above water in dense vegetation. A nocturnal feeder on aquatic plants, this exotic species can seriously damage wetlands and crops. Introduced from South America for fur farming in Louisiana and Oregon; now widespread in marshes and lakes in much of the south and northwestern United States.

AMERICAN BEAVER *Castor canadensis* 100–120cm, 23–32cm, 16–30kg

Unmistakable large, aquatic rodent with a sizeable, flat, paddle-shaped, scaly tail. Our largest rodent. Pelage is brown with shiny guard hairs and grayish underfur. Hind feet are webbed. Incisor teeth are large, orange, and ever-growing. Distinctive flattened tail is used as a rudder; also slapped against the top of the water as an alarm. Eats the leaves and inner bark of many tree species, preferring willow and aspen. Survives long, harsh winters by huddling in its insulated lodge, storing fat in its tail, and retrieving and eating underwater food caches. Lives in small family groups. The presence of a Beaver family in an area is easily detected by the saplings and small trees they cut down, strip of bark, and use to build dams and dome-shaped lodges. Typically nocturnal, most often seen around dawn or dusk. Once trapped to extinction in many areas, the Beaver has come back, and is now common in many areas, sometimes considered a pest. Lives in a variety of rivers and lakes.

ROUND-TAILED MUSKRAT

MUSKRAT

NUTRIA

AMERICAN BEAVER

PLATE 16
MARMOTS

MARMOTS – These large, chunky squirrels have short stubby tails. Because of their terrestrial and diurnal habits, they can be easy to observe. Often leave feces on rocks or logs. All retreat to their burrows for protection, and to hibernate. These vegetarians are typically found near open, grassy habitats.

OLYMPIC MARMOT *Marmota olympus* 68–78cm, 19–25cm, 5–7kg

A large, drab brown marmot from the Olympic Peninsula, Washington (yellow area on map). Browner than the Hoary Marmot, but otherwise similar in having a long tail, white nose, and white band in front of the eyes. Coat bleaches yellowish in the summer. Males are larger. Typically hibernates from September to May or June. Prefers montane slopes with rock talus and lush meadows between 1700 and 2000m.

HOARY MARMOT *Marmota caligata* 62–85cm, 17–25cm, 5–6kg

A large gray marmot with a relatively long tail. Hairs on the rump are tinged buff. The tail is brownish. Face is marked with white in front of the eyes and a dark band on snout. Dark streaks also mark the side of the head and neck. Typically hibernates from September to May. Alpine-montane specialist except in Alaska where it ranges down to sea level. Lives in treeless meadows where rocky outcrops and talus provide burrows.

YELLOW-BELLIED MARMOT *Marmota flaviventris* ♂ 49–70cm, 15–22cm, 3–5kg; ♀ 47–67cm, 13–22cm, 1.6–4.0kg

A small gray marmot with yellow on the belly and neck. There is a white band across the nose. Top of head is black. Hibernates to avoid cold and snow; timing of hibernation depends on age, sex, and local weather. Uses meadows adjacent to talus slopes or rock outcrops.

ALASKA MARMOT *Marmota broweri* ♂ 58–65cm, 15–18cm, 3.0–4.0kg; ♀ 54–60cm, 13–16cm, 2.5–3.5kg

A dark marmot with a black nose and top of the head. Tail is short. Typically hibernates September to June. Only marmot in the Brooks mountain range. Found near boulder fields and talus slopes.

VANCOUVER ISLAND MARMOT *Marmota vancouverensis* 58–75cm, 16–30cm, 3.0–6.5kg

A dark marmot from British Columbia (see red area on map of Olympic Marmot) with white on the face, chest and belly. Males are larger. Typically hibernates October to April or May. Endangered because of restricted range, most colonies within an 80km². Lives in alpine and subalpine meadows from 1000 to 1460m altitude.

WOODCHUCK *Marmota monax* 41–67cm, 10–15cm, 3–4kg

A grizzled grayish or brown marmot with reddish-brown underparts. Feet are typically blackish brown, but are pinkish in Alaska. The subspecies *M. m. ochracea* from northwest Canada and Alaska is reddish cinnamon in color. Males are larger. Widespread and common in meadows and along forest edges.

HOARY MARMOT

OLYMPIC
MARMOT

feces

ALASKA MARMOT

YELLOW-BELLIED
MARMOT

VANCOUVER
ISLAND
MARMOT

typical

WOODCHUCK

far north
Woodchuck

PLATE 17
EASTERN AND TROPICAL
TREE SQUIRRELS

TREE SQUIRRELS – These familiar diurnal squirrels are grouped by their large bushy tails and tree-climbing habits. Many species have a variety of color morphs.

EASTERN GRAY SQUIRREL *Sciurus carolinensis* 38–52cm, 15–24cm, 338–750g

A gray squirrel with a bushy tail edged in white. Belly is whitish. Gray back may have a red-brown tinge. Black-morph individuals and albinos can be quite common in some areas. Makes leaf nests the size of bushel-baskets. The most commonly seen mammal in the eastern United States. Introduced in many western cities. Favors hardwood or mixed forests, including residential areas.

EASTERN FOX SQUIRREL *Sciurus niger* 45–70cm, 20–33cm, 696–1233g

A large squirrel with a bushy tail edged with brown or orangish brown. Usually twice the size of the Eastern Gray Squirrel, with a more colorful coat and a brownish tinge to the tail. The most typical color phase has rusty-gray upperparts with a rusty- yellow or orange belly. Other color morphs include an all-black form, a southeastern form that is black or dark brown with a white nose and ears, and a gray form with rusty limbs, a black head and a white nose and ears. Makes leaf nests like the Eastern Gray Squirrel. Introduced to some parts of California, Colorado, Oregon, and Washington. Generally abundant, although the subspecies from the eastern shore of Maryland is considered endangered. These savanna animals prefer open, parklike habitats with scattered trees and an open understory.

RED-BELLIED SQUIRREL *Sciurus aureogaster* 42–57cm, 21–31cm, 375–680g

A colorful squirrel introduced to the Florida Keys in 1938. It is either gray or frosted with white. The underside and sides up to its shoulders are mahogany red. Tail is mixed with black and white. Throughout its range in Latin America, this squirrel is variable in coloration, and black melanistic animals are common. Shy denizen of the treetops. In North America, known only from wooded areas on Elliott, Sand, and Adams Keys.

MEXICAN FOX SQUIRREL *Sciurus nayaritensis* 49–61cm, 24–30cm, 628–814g

A unique, vividly colored, reddish or orangish squirrel. Dark tail is incredibly bushy and edged in white. Winter pelage has a broad band of blackish hair running down the back; body is yellowish gray, and belly, feet and eye ring are orangish yellow. In summer the back is a grizzled mixture of pale orangish yellow and black; underparts are tawny. Inhabits partially open pine-oak forests.

black morph

typical color

leaf
nest

EASTERN
GRAY
SQUIRREL

EASTERN FOX
SQUIRREL

southeastern
form

typical
color

black morph

black-headed
morph of Eastern
Fox Squirrel

RED-BELLIED
SQUIRREL

winter

MEXICAN FOX SQUIRREL

summer

PLATE 18
WESTERN TREE SQUIRRELS

ABERT'S SQUIRREL *Sciurus aberti* 64–58cm, 19–28cm, 540–971g

This squirrel has tufted ears and white on the underside of a broad tail. New ear tufts are grown in October, and may be inconspicuous by summer. The back is gray, with a rusty band. Some animals in central Colorado are uniform brown. North of Grand Canyon, the subspecies *S. a. kaibabensis* has dark underparts and all-white tail, and was once considered a distinct species. Elsewhere the tail is only white on the underside and the belly is white. The back and belly color are usually separated by a black line. There is also a dark melanistic form. Lives chiefly in ponderosa pine forests.

ARIZONA GRAY SQUIRREL *Sciurus arizonensis* 45–57cm, 20–31cm, 527–884g

A gray squirrel with a black tail that is washed in white above and marked with an orange or rusty-brown center stripe below. Back and sides are salt-and-peppered steel gray, often mottled with brown or rusty yellow. Underparts are white. Uncommon and typically shy in riparian broadleaf forests of Arizona, New Mexico, and Mexico.

WESTERN GRAY SQUIRREL *Sciurus griseus* 51–77cm, 24–38cm, 500–950g

This gray squirrel has a long pepper-gray tail with white edging. Silver-gray back contrasts with pure white belly. A patch behind the ears is pale reddish brown. Feet are dusky. Pelage is more silvery and ears are relatively larger than in the Eastern Gray Squirrel. Shy and intolerant of humans. Threatened in the north of its range. Uses oak-conifer woodlands.

EASTERN GRAY SQUIRREL *Sciurus carolinensis*

Introduced into some residential areas in the west. Tail is typically browner than the Western Gray Squirrel, lacking any silvery tint. (*See* page 48 for more details.)

ABERT'S
SQUIRREL

typical
winter

black morph
winter

typical
summer

Colorado
brown
morph of
Abert's
Squirrel

ARIZONA
GRAY
SQUIRREL

winter "Kaibab" morph of
Abert's Squirrel

WESTERN
GRAY
SQUIRREL

EASTERN
GRAY
SQUIRREL

PLATE 19
RED AND FLYING SQUIRRELS

TAMIASCIURUS – Small red, diurnal tree squirrels. Discard pine cones to form middens under favorite perches. Retire in tree holes and ball nests. Give cicada-like buzz call in spring; territorial bark and chatter year-round. A third species lives in Baja California, Mexico.

DOUGLAS'S SQUIRREL *Tamiasciurus douglasii* 270–350mm, 100–156mm, 141–312g

Less reddish than the Red Squirrel with a gray to orange eye ring and belly. Brownish or reddish-gray sides; there is a chestnut-brown band down middle of back. Black ear tufts and side stripe are more prominent in winter. Uses coniferous and mixed forests.

RED SQUIRREL *Tamiasciurus hudsonicus* 270–380mm, 90–150mm, 140–250g

Redder than Douglas's Squirrel with a white eye ring and belly. In summer a distinct black stripe forms between the rusty olive-brown back and white belly. In winter this stripe fades, but other colors are brighter; feet and ventral surface become more gray. Mt. Graham, Arizona, subspecies is Endangered. Common in coniferous forests and southern mixed forests.

FLYING SQUIRRELS – Unique with their large, black eyes and loose flap of skin along sides of body that allows them to glide (not fly) between trees. The two species are separated by geographic range and color of the base of the belly fur. Both have a broad flattened tail that serves as a stabilizer and rudder, as well as a cartilaginous rod extending from the side of their wrist to steer their glides. Both are nocturnal and spend most of their time in the trees. They emit high-pitched squeaks and twitters, nest in tree holes, vine tangles, and nest boxes, sleep in groups in the winter, and eat nuts, berries, fungus, insects, and eggs.

SOUTHERN FLYING SQUIRREL *Glaucomys volans* 120–140mm, 80–120mm, 46–85g

Hair on the belly is white at both the base and the tip. Back is brown, grayish or tawny; edge of the wing membrane is blackish. Common in hardwood forests, less so in mixed and coniferous forests.

NORTHERN FLYING SQUIRREL *Glaucomys sabrinus* 190–300mm, 90–140mm, 38–123g

Hairs on belly are white at tips but dark gray at base. Otherwise like Southern Flying Squirrel except larger size. Often has pale patches of fur at the base of ears and a dark tail tip. Known to visit feeding stations. Often common, but Endangered in North Carolina and Virginia. Most common near water in coniferous forests, less in mixed and pure hardwood forests.

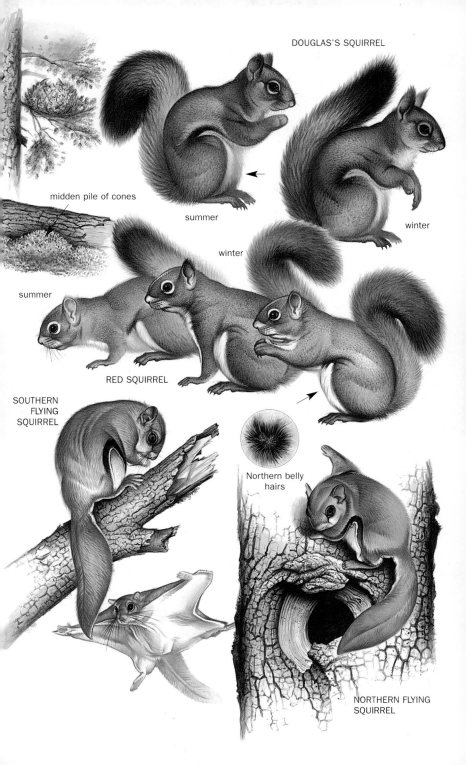

DOUGLAS'S SQUIRREL

midden pile of cones

summer

winter

winter

summer

RED SQUIRREL

SOUTHERN
FLYING
SQUIRREL

Northern belly
hairs

NORTHERN FLYING
SQUIRREL

PLATE 20
PRAIRIE DOGS

PRAIRIE DOGS – These sandy-colored squirrels live in extensive colonies of underground burrows. Conspicuous with their diurnal feeding and barked alarm calls. All species are vulnerable to human persecution because of an exaggerated and often unwarranted reputation as pests. The four species are distinguished by the color of their back and tail tip.

GUNNISON'S PRAIRIE DOG *Cynomys gunnisoni* ♂ 320–390mm, 40–60mm, 460–1300g; ♀ 310–340mm, 50–60mm, 465–750g

Has a gray tail tip with a white border. Head and back are sandy-colored, belly is white. Lives in colonies that are smaller than other prairie dogs, with less modified mounds and more vegetation between holes. Contact call between individuals is a raspy chatter. Hibernates from October or November until March or April. Lives in montane valleys and high plateaus in the southern Rockies.

WHITE-TAILED PRAIRIE DOG *Cynomys leucurus* ♂ 350–390mm, 40–70mm, 850–1650g; ♀ 320–370mm, 50–60mm, 705–1050g

A buff or gray-colored prairie dog with a white tail tip. Has a dark brown spot above the eye and on the cheek. Contact call between individuals is a laughing bark. Typically hibernates from October to April. Lives in mountain meadows. May venture into semidesert areas in the north of the range.

UTAH PRAIRIE DOG *Cynomys parvidens* ♂ 300–370mm, 50–60mm, 460–1250g; ♀ 290–370mm, 50–60mm, 410–790g

A reddish or cinnamon-colored prairie dog with a white tail tip. Burrow entrances are marked by a mound of dirt. Hibernates in winter. Endangered because widespread human persecution has driven them out of 90% of their historic range. Now limited to grasslands and flat plains in southern Utah.

BLACK-TAILED PRAIRIE DOG *Cynomys ludovicianus* ♂ 360–430mm, 70–90mm, 575–1490g; ♀ 340–400mm, 60–90mm, 765–1030g

A large prairie dog with a black tail tip. Contact call between individuals is a "we-oo." Burrow entrances are surrounded by a large mound of dirt shaped into domes. Dormant in winter, but does not truly hibernate, and can be seen above ground on warm winter days. Clips vegetation very short around colonies for an unobstructed view. Lives at high densities where not persecuted. Prefers shortgrass to mid-grass prairies.

GUNNISON'S PRAIRIE DOG

WHITE-TAILED
PRAIRIE DOG

UTAH
PRAIRIE
DOG

BLACK-TAILED
PRAIRIE DOG

PLATE 21
LARGE, SPOTTED GROUND SQUIRRELS

LARGE SPOTTED GROUND SQUIRRELS – These four large, spotted ground squirrels are often conspicuous with their diurnal habits. All put on considerable weight leading up to the late summer or fall start of their hibernation. Coloration and tail size are the key to distinguishing the different species.

ROCK SQUIRREL *Spermophilus variegatus* 470–500mm, 190–230mm, 450–875g

Has the bushiest tail of any ground squirrel. Back is typically a buff gray, and is marked with flecks of white; this spotting sometimes producing a wavy striped appearance, but never the white shoulder markings of *S. beecheyi*. Often has varying amounts of black on the back; this may be across the head, shoulders, back, or entire body. Head varies from pinkish buff to brown. Has conspicuous white crescents above and below the eyes. Tail not as bushy as tree squirrels. Rarely climbs trees. Males are larger. Hibernation is short and intermittent. Typically colonial. Lives in rocky canyons, cliffs and hillsides.

CALIFORNIA GROUND SQUIRREL *Spermophilus beecheyi* 360–500mm, 140–230mm, 350–885g

A large, spotted squirrel with a mantle of light gray running from the ears back onto the shoulders. This gray streak is inconspicuous or absent in some subspecies. Back is brown; dorsal buff flecking is heaviest on rump and flanks. Belly is pale gray. Bushy tail is dark gray above and paler below. Hibernates in winter, timing depends on local weather. Common in successional habitat including roadsides, chaparral, and open grassy areas.

COLUMBIAN GROUND SQUIRREL *Spermophilus columbianus* 320–410mm, 80–120mm, 340–812g

A large ground squirrel with a reddish-brown nose and forelegs. Back is grayish with indistinct buff spotting. Bushy tail is reddish and edged with white and some black hairs. Hibernates for more than two thirds of the year. Lives in colonies in alpine meadows and grassy lowlands.

ARCTIC GROUND SQUIRREL *Spermophilus parryii* 330–490mm, 80–150mm, 530–816g

Unique with reddish-brown back that is flecked with whitish spots. Spring–summer pelage is reddish-brown while fall–winter fur is grayish, with a cinnamon-brown head color. Males are larger. Both sexes get much heavier just before hibernation. Hibernates for seven winter months. Alarm calls include a shrill whistle or sharp "cheek-chick" call. Found in tundra and mountain meadows above the timberline.

variable amount
of black

ROCK
SQUIRREL

CALIFORNIA
GROUND
SQUIRREL

COLUMBIAN
GROUND
SQUIRREL

early summer

late summer

ARCTIC GROUND SQUIRREL

PLATE 22
BUSHY-TAILED, FLECKED GROUND SQUIRRELS

BUSHY-TAILED, FLECKED GROUND SQUIRRELS – These large ground squirrels have pale flecks on their back that do not form distinct spots. The color and size of their tail distinguishes the different species. All are diurnal seed and leaf eaters. These burrow-living species put on considerable weight leading up to the late summer–early fall onset of hibernation.

FRANKLIN'S GROUND SQUIRREL *Spermophilus franklinii*
355–410mm, 120–158mm, 340–950g

A large ground squirrel with a gray head and a long, gray, bushy tail. Underparts dull whitish or buff. Body and ears are smaller than the tree squirrels. Gives loud whistle alarms and musical trills. Hibernates August–September until April–May. Colonies in a variety of habitats including tall grass, shrub land, and woodland edge.

BELDING'S GROUND SQUIRREL *Spermophilus beldingi*
♂ 270–315mm, 60–75mm, 300–450g; ♀ 265–295mm, 60–75mm, 230–400g

Tail is moderately bushy, reddish, and black-tipped. There is a broad, brown band that runs down the center of the back, contrasting slightly with the grayish sides. Has a narrow, white eye ring. Extensive hibernator, so that all foraging, growth and reproduction occurs in three frantic spring and summer months. Typical alarm call is a trill of 5–8 short whistles. Uses short grass habitats in alpine meadows, along roadsides, and in cultivated fields.

UINTA GROUND SQUIRREL *Spermophilus armatus* 270–320mm,
43–81mm, 250–600g

Uniform speckled brownish or grayish back with a grayish undertail. Head and neck are always gray, back varies from grayish to brownish to cinnamon buff. Not as buff as the Wyoming Ground Squirrel, with gray (not buff) under the tail. Larger than the Piute Ground Squirrel, with a longer tail. Gives a variety of calls including chips, churrs, and squeals. Use sagebrush, grassy meadows, and sometimes cultivated lawns.

WYOMING GROUND SQUIRREL *Spermophilus elegans* 253–307mm,
59–79mm, 286–411g

A flecked, drab-colored ground squirrel with a buff undertail. Flecks may be pale pinkish buff, clay-colored, or pinkish cinnamon. Underparts are cinnamon and gray-colored. Distinguished from the similar Belding's and Uinta Ground Squirrels by having a buff (not reddish or grayish) underside of the tail with a white edging. Churr and chip calls are higher pitched than in Richardson's Ground Squirrel. Uses mountain meadows above 1500m and talus slopes above the timberline.

RICHARDSON'S GROUND SQUIRREL *Spermophilus richardsonii*
♂ 283–337mm, 65–88mm, 290–745g; ♀ 264–318mm, 55–82mm, 120–590g

A flecked, drab ground squirrel with a white-edged tail. Gray fur is washed with cinnamon buff above and pale buff or white below. Tail is bordered by white, and clay-colored or light brown below. Distinguished from the similar Belding's and Uinta Ground Squirrels by having a brownish (not reddish or grayish) underside of the tail with a white edging. Churr and chip calls are lower pitched than in Wyoming Ground Squirrel. Hibernates from September to March.

BELDING'S GROUND
SQUIRREL

FRANKLIN'S GROUND
SQUIRREL

UINTA GROUND SQUIRREL

WYOMING
GROUND
SQUIRREL

RICHARDSON'S
GROUND SQUIRREL

PLATE 23
THIN-TAILED, UNMARKED
GROUND SQUIRRELS

THIN-TAILED, UNMARKED GROUND SQUIRRELS – Small squirrels with plain backs without flecks or spotting. Color and size of their thin tails distinguish species. Like other ground squirrels, they are diurnal leaf and grain eaters that retire to burrow systems.

ROUND-TAILED GROUND SQUIRREL *Spermophilus tereticaudus*
202–278mm, 60–112mm, 110–170g

This small squirrel has whitish cheeks, and a darker top of the head. Undertail is buff or cinnamon. Pelage is plain, and two distinct color morphs are known with the back being either cinnamon or drab gray-brown. Sides of head are dull white. Belly is white. Smaller than the Mohave Ground Squirrel with white cheeks and a buff undertail. Has a thinner tail than other ground squirrels, and no spots or flecking. Does not hibernate, but may enter torpor in the winter. Call with high-pitched whistles or peeps. Uses sandy flat areas in the Sonoran and Mohave deserts.

MOHAVE GROUND SQUIRREL *Spermophilus mohavensis*
210–230mm, 57–72mm, 70–300g

A small spotless squirrel with brownish cheeks and a white undertail. Tail has short hairs and is creamy white underneath. Larger than the Round-tailed Ground Squirrel with a shorter tail that has a white (not buff or cinnamon) undersurface, and brown cheeks. Has a thinner tail than other ground squirrels, and no spots or flecking. Often holds tail over back to display the creamy white underside. Hibernates in winter. Uses level sandy areas with sparse shrub growth in the Mohave desert.

SPERMOPHILUS TOWNSENDII SPECIES GROUP – Small squirrels with short ears and unmarked pelage. Color is grayish washed with pinkish buff above and buff-white below. The sides of the head and hind legs are tinted with reddish buff. Underside of tail is light cinnamon. In addition to geographic range, they can be distinguished from other ground squirrels by their lack of flecking, and thin tail. These three species were originally considered conspecific, and were only recently split by genetic studies. There are no known morphological differences to distinguish them.

TOWNSEND'S GROUND SQUIRREL *Spermophilus townsendii*
200–232mm, 39–54mm, 125–325g

Has 36 chromosomes. Hibernates from May–June till January–February. Uses sagebrush and agricultural areas north of the Columbia River and south of the Yakima River in south central Washington (red area on map).

MERRIAM'S GROUND SQUIRREL *Spermophilus canus* 190–217mm, 37–42mm, 100–250g

Has 46 chromosomes. Hibernates from August till March. Uses grasslands and pastures with big sagebrush and western juniper (see yellow area on map of Townsend's Ground Squirrel for distribution).

PIUTE GROUND SQUIRREL *Spermophilus mollis* 201–233mm, 44–61mm, 84–205g

Has 38 chromosomes. Hibernates or sleeps for most of winter. Uses agricultural areas and desert communities.

gray-brown morph

ROUND-TAILED
GROUND SQUIRREL

cinnamon
morph

MOHAVE
GROUND
SQUIRREL

TOWNSEND'S, MERRIAM'S
AND PIUTE GROUND SQUIRRELS

PLATE 24
SMALL, SPOTTED GROUND SQUIRRELS

SMALL, SPOTTED GROUND SQUIRRELS – These five small ground squirrels all have spots that are more distinct than the diffuse speckling sometimes found in "unspotted" ground squirrels. All are diurnal. Typically leaf and seed eaters, most will also eat insects or meat when they are available.

THIRTEEN-LINED GROUND SQUIRREL *Spermophilus tridecemlineatus* 170–310mm, 60–132mm, 110–270g

Back is marked with 13 alternating stripes that are either solid pale or dashed and dark. Tail and back are mixed brown and white. Has a white eye ring. Heaviest just before hibernation in September and October. Emerges from hibernation in March or April. Rarely ventures far from escape burrow. Warning call is a soft, trilled whistle. Often seen standing upright on a roadside or other mowed area

MEXICAN GROUND SQUIRREL *Spermophilus mexicanus* 280–380mm, 110–166mm, 137–330g

Has nine rows of squarish white spots on its brown back. Larger than *S. tridecemlineatus* with no solid white stripes. Belly is whitish or buff. Tail is moderately bushy. Head is brown, tip of nose is often cinnamon or yellowish. Has a white eye ring. Males are larger. Northern populations hibernate. Lives alone or in colonies in grassy habitats or arid areas with brushy vegetation.

SPOTTED GROUND SQUIRREL *Spermophilus spilosoma* 185–253mm, 55–92mm, 100–200g

A small, drab ground squirrel with light spotting and a bushy tail. Some populations are more spotted than others. Short round tail is dark, with a black tip and a cinnamon underside. Belly is light-colored. Color of back is variable and blends in with the local substrate including: cinnamon drab, smoke gray, and brownish morphs. Northern populations hibernate from July–Sept to April. Use desert scrublands and grasslands.

WASHINGTON GROUND SQUIRREL *Spermophilus washingtoni* 185–245mm, 32–65mm, 120–300g

A small, brownish-gray squirrel with distinct whitish spots. Found only in Washington and Oregon (yellow area on map). Grayish-white color on the belly and sides changes abruptly to the darker dorsal color, forming a lateral line between the two colors. Underside of tail, nose, and lower legs are pinkish cinnamon (not rufous as in the Idaho Ground Squirrel). Also has smaller ears (10–13mm). Extensive hibernators, typically only active above ground from late February to early summer. Threatened by hunting and habitat destruction. Uses dry open sagebrush or grassland habitat.

IDAHO GROUND SQUIRREL *Spermophilus brunneus* 209–258mm, 39–62mm, 120–290g

A small, lightly spotted squirrel with russet-colored legs, nose, and tail. Only found in Idaho (see red area on map of Washington Ground Squirrel). The two known subspecies can be distinguished, and may actually be different species. *S. b. brunneus* from Adams and Valley counties is reddish gray with a buff-white eye ring. *S. b. endemicus*, from Gem, Payette, and Washington counties is grayish brown with a creamy white eye ring and russet on the legs and base of tail. Lives in burrow systems dug in select mountain meadows.

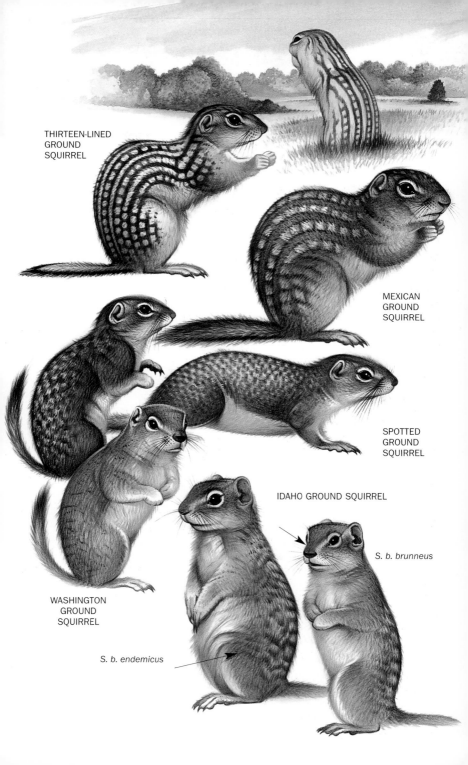

THIRTEEN-LINED
GROUND
SQUIRREL

MEXICAN
GROUND
SQUIRREL

SPOTTED
GROUND
SQUIRREL

IDAHO GROUND SQUIRREL

S. b. brunneus

WASHINGTON
GROUND
SQUIRREL

S. b. endemicus

PLATE 25
STRIPY GROUND SQUIRRELS

ANTELOPE SQUIRRELS – These small, striped squirrels are recognized by their single pair of white stripes that run from shoulder to rump. Unlike chipmunks, they have no facial stripes. They often carry their tail arched forward over their back. All are diurnal and call with long, high-pitched trills.

HARRIS'S ANTELOPE SQUIRREL *Ammospermophilus harrisii*
216–267mm, 67–92mm, 113–150g

Underside of tail is grayish from mixing of black and white hairs. Back is pale brown to blackish. Eye ring and belly are whitish. Often sits erect on hind feet. Runs with tail held vertically. Burrows under shrubs in a variety of desert habitats.

TEXAS ANTELOPE SQUIRREL *Ammospermophilus interpres*
220–235mm, 68–84mm, 99–122g

Underside of tail is white with two dark bands. Back and basal third of tail is gray. Last two thirds of tail is grayish black. Shoulders, hips, and outer legs are yellowish brown. Belly is white. Uses rocky and shrubby areas around desert mountain ranges.

WHITE-TAILED ANTELOPE SQUIRREL
Ammospermophilus leucurus 188–239mm, 42–87mm, 96–117g

Underside of tail is white with one dark band. Subspecies vary from grayish brown to cinnamon. Fur is longer and grayer in winter. Uses a variety of shrubby desert habitats.

NELSON'S ANTELOPE SQUIRREL *Ammospermophilus nelsoni*
230–267mm, 67–78mm, 142–179g

Coloration is generally buff rather than gray. Underside of tail is creamy white. Upperparts are a dull yellow-brown and underparts and eye ring are whitish. Only antelope squirrel in range. Inhabits gentle slopes with shrubby cover.

GOLDEN-MANTLED GROUND SQUIRRELS – Their conspicuous light and dark side stripes and colorful heads identify these two small ground squirrels. They are larger than chipmunks and lack face stripes. The two species are distinguished by range and the distinctiveness of their colorful hood.

GOLDEN-MANTLED GROUND SQUIRREL *Spermophilus lateralis*
355–410mm, 120–158mm, 340–425g

A small ground squirrel with a dorsal stripe and well defined golden hood. Head and neck may be golden brown, tawny, or russet-colored. Winter pelage is grayer. Does not overlap (yellow area on map) with the larger Cascade Golden-mantled Ground Squirrel.

CASCADE GOLDEN-MANTLED GROUND SQUIRREL *Spermophilus saturatus*
286–315mm, 92–118mm, 200–300g

A small ground squirrel with a dorsal stripe and poorly defined russet hood. Sometimes considered a subspecies of Golden-mantled Ground Squirrel (see red area on map of Golden-mantled for distribution).

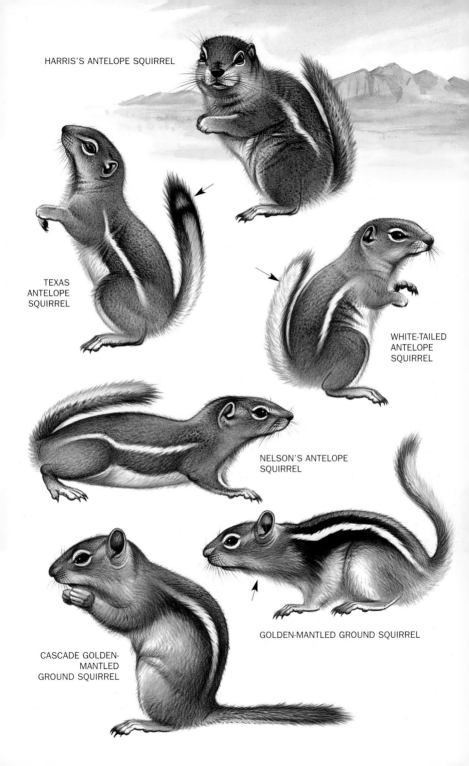

HARRIS'S ANTELOPE SQUIRREL

TEXAS
ANTELOPE
SQUIRREL

WHITE-TAILED
ANTELOPE
SQUIRREL

NELSON'S ANTELOPE
SQUIRREL

GOLDEN-MANTLED GROUND SQUIRREL

CASCADE GOLDEN-
MANTLED
GROUND SQUIRREL

PLATE 26
EASTERN AND ROCKY MOUNTAIN CHIPMUNKS

CHIPMUNKS – This group of small, striped squirrels is easy to recognize, but it is very difficult to distinguish between the 22 different species. Easterners have it easy, with only one to choose from. Western naturalists should consider geographic range and subtle color differences. The Eastern Chipmunk hibernates in the winter while western species rely on underground stores of fruit and seed food to survive the winter.

LEAST CHIPMUNK *Tamias minimus* 185–216mm, 78–113mm, 32–50g

A very small chipmunk. The subspecies *T. m. sylvaticus* shows coloration typical of many central and northern forms. The subspecies *T. m. scrutator* from far western United States has gray on the rump, central dorsal stripe, and head. Southwestern forms are paler brown (see page 68). Runs with tail held vertically. Typically flicks tail up and down when perched. Calls short and infrequent. Most abundant in alpine coniferous forests and tundra, also found in aspen woodlands, meadows, scrublands, and sagebrush desert.

EASTERN CHIPMUNK *Tamias striatus* 215–285mm, 80–115mm, 80–150g

This red-rumped rodent is the only chipmunk in most of eastern North America. Our largest chipmunk. Color of back varies from pale brown in southern Ontario, to dark red Appalachian forms, and brightly colored animals in the southwest of their range. Vocalizes with a series of chips and a high-pitched alarm call. Common in eastern deciduous forests.

CLIFF CHIPMUNK *Tamias dorsalis*

T. d. utahensis from Utah has a more distinct dark stripe down the center of its back than other subspecies. Identifiable by pale gray color of back. (For more details, see page 68.)

RED-TAILED CHIPMUNK *Tamias ruficaudus* 223–248mm, 100–121mm, 53–62g

Has a long tail that is red on the underside. Winter form is paler, but still has red tail. Some suggest that the subspecies *T. r. simulans* from southern British Columbia, northeast Washington, north Idaho, and northwest Montana should be elevated to full species based on the morphology of their genital bones. Common in dense coniferous forests.

YELLOW-PINE CHIPMUNK *Tamias amoenus* 186–238mm, 72–109mm, 36–50g

A mid-sized, yellowish chipmunk. Larger than the Alpine and Least Chipmunk, smaller than other chipmunks. Has brighter colors than the Least Chipmunk, with a reddish (not yellowish) undertail. Common in chaparral, meadows, and rocky outcrops.

UINTA CHIPMUNK *Tamias umbrinus* 210–240mm, 84–119mm, 51–74g

A brownish chipmunk with brown and white dorsal stripes. Difficult to distinguish by sight. Some consider it conspecific with Palmer's Chipmunk. Has less red than the Panamint Chipmunk and is larger than the Least or Yellow-pine Chipmunks. Holds tail horizontal when running. Often climbs trees. Has five-syllable call. Common in montane coniferous forests, especially above 1830m.

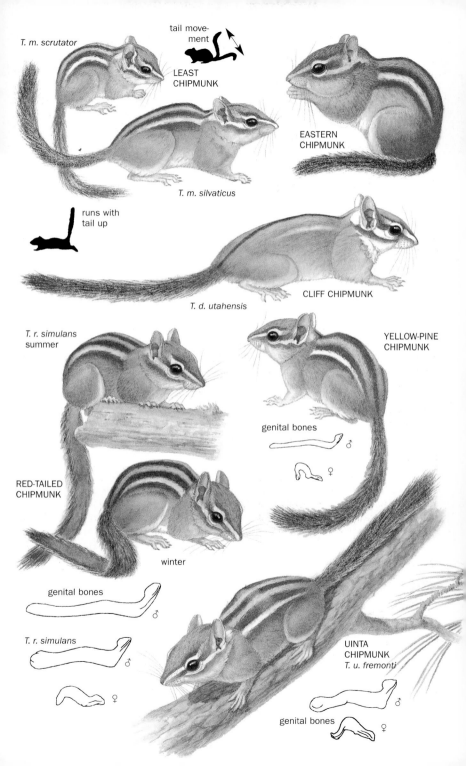

T. m. scrutator

tail movement

LEAST CHIPMUNK

EASTERN CHIPMUNK

T. m. silvaticus

runs with tail up

T. d. utahensis

CLIFF CHIPMUNK

T. r. simulans summer

YELLOW-PINE CHIPMUNK

genital bones ♂

♀

RED-TAILED CHIPMUNK

winter

genital bones ♂

T. r. simulans ♂

♀

UINTA CHIPMUNK *T. u. fremonti*

genital bones ♂

♀

PLATE 27
SOUTHWEST CHIPMUNKS

GRAY-COLLARED CHIPMUNK *Tamias cinereicollis* 208–242mm, 90–109mm, 55–70g

Prominently marked with gray on the cheeks, shoulders and rump. Dorsal stripes are darker than those of the Gray-footed Chipmunk. Hind feet are pinkish buff. In winter the flanks and tail edges are paler, and the belly has more gray. Runs with tail held horizontally. Often sits on logs slowly waving tail from side to side while "chucking." Alarm call is a shrill "chipper." Climbs trees more than other chipmunks. Has a limited range in montane coniferous forests between 1950 and 3440m.

GRAY-FOOTED CHIPMUNK *Tamias canipes* 210–264mm, 91–115mm, 65–75g

Hind feet are gray on the dorsal surface. Stripes on back are a dark rusty color (not with a solid black center like the Gray-collared Chipmunk). Has a grayer head than the Colorado Chipmunk, with gray (not orange) shoulders. Females are slightly larger. Vocalizes from hiding spots with a light "chipper" or a low repeated "chuck." Most active at dawn on rocky slopes with thick brush.

CLIFF CHIPMUNK *Tamias dorsalis* ♂ 204–226mm, 82–100mm, 54–64g; ♀ 212–235mm, 89–105mm, 58–67g

Gray back has indistinct stripes. Center dorsal stripe dark, others faint. White patch behind ears. Sides are brownish, underparts are creamy white. Tail is brownish orange below. Call is a series of chips with a terminal pulse. Common in a variety of habitats with cliffs or boulders. (For Utah form, see page 66.)

LEAST CHIPMUNK *Tamias minimus*

The southwestern subspecies are paler than those in the Rockies. Distinguished from other southwestern chipmunks by small size, dull color, and lack of gray. Runs with tail held vertically, often flicks tail up and down. (For more details, see page 66.)

COLORADO CHIPMUNK *Tamias quadrivittatus* 212–245mm, 80–118mm, 54–80g

Has bright, contrasting stripes, a grayish forehead, and orangish shoulders. *T. q. australis* from southern Arizona has some gray on shoulders, but is distinguished from the Gray-collared Chipmunk by being smaller and having an orange wash on shoulder. *T. q. quadrivittatus* from northern Arizona, Utah, and Colorado has brighter orange flanks than the similar Uinta Chipmunk, with more red on the tail underside and darker dorsal stripes that extend all the way to the base of the tail. Uses rocky slopes in Ponderosa pine forests.

HOPI CHIPMUNK *Tamias rufus* 197–221mm, 81–95mm, 52–62g

Stripes are tawny and underside of tail is reddish. Where they overlap, the larger Colorado Chipmunk has darker stripes, and the smaller Least Chipmunk has browner stripes. Tail is carried horizontally when running. Uses rocky areas in pinyon and juniper woodlands.

UINTA CHIPMUNK *Tamias umbrinus*

T. u. adsitus from the Kaibab Plateau is similar to the sympatric Hopi Chipmunk, but has darker dorsal stripes. (For more details, see page 66.)

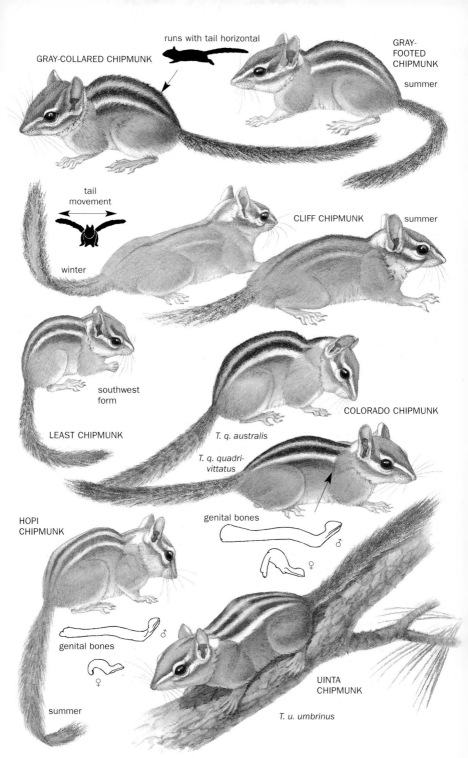

GRAY-COLLARED CHIPMUNK

runs with tail horizontal

GRAY-FOOTED CHIPMUNK

summer

tail movement

winter

CLIFF CHIPMUNK

summer

LEAST CHIPMUNK

southwest form

COLORADO CHIPMUNK

T. q. australis

T. q. quadri-vittatus

HOPI CHIPMUNK

genital bones

♂
♀

genital bones

♂
♀

summer

UINTA CHIPMUNK

T. u. umbrinus

PLATE 28
SOUTHERN CALIFORNIA CHIPMUNKS

LODGEPOLE CHIPMUNK *Tamias speciosus* ♂ 220–222mm, 79–100mm, 51–61g; ♀ 197–229mm, 67–102mm, 55–69g

A brightly colored chipmunk with broad, pure white outer dorsal stripes. Outermost dark stripes are poorly distinguished. Underside of tail has a black tip. Sides are reddish brown. Top of head and rump are grayish. Winter pelage is grayer overall, with browner dark dorsal stripes. Larger than the Yellow-pine Chipmunk, with less distinct outermost dark stripes. Smaller than the Long-eared Chipmunk, with relatively smaller ears. Climbs trees more than most chipmunks. Common in open coniferous forests.

MERRIAM'S CHIPMUNK *Tamias merriami* 240–255mm, 100–115mm, 70–80g

Large chipmunk with a bushy tail. Contrasting dorsal stripes are light gray and brown. Duller, but very similar to California Chipmunk; genital bones may be needed for certain ID. Distinguished from Townsend's Chipmunk by having gray (not brown) cheeks. Edging on tail is buff-colored, not white like in the Long-eared Chipmunk. Darker on coast (see page 72). Calls with "chips" and a trill from elevated perches in pine and oak forests.

CALIFORNIA CHIPMUNK *Tamias obscurus* 200–250mm, 70–120mm, 60–84g

Large chipmunk with a bushy tail. Head is grayish on top and striped in white on the sides. Back has brown and gray stripes. Striping is less distinguished in winter pelage. Overlaps slightly in range with the very similar Merriam's Chipmunk; genital bones needed for certain ID. Stands on hind feet to give "chip" vocalization. Common in sandy and rocky areas with sparse pinyon pine, juniper, and scrub oak vegetation (yellow area on map).

PALMER'S CHIPMUNK *Tamias palmeri* 210–223mm, 87–101mm, 50–69g

A small chipmunk from the Spring Mountains of southwest Nevada. Has only three (not the usual five) dark stripes on back. The center pair of pale stripes is grayish, the outer pair and belly are creamy white. Colors are more muted in the winter. Only the larger Panamint Chipmunk overlaps in range. These two are also distinguished by the grayer shoulders of Palmer's Chipmunk. May be conspecific with *T. umbrinus*; genital bones are shown for certain ID. Emits chippering, chip, and chuck calls along cliffs. Found in rocky coniferous habitat above 2000m elevation (see red area on map of California Chipmunk).

PANAMINT CHIPMUNK *Tamias panamintinus* ♂ 230–239mm, 85–100mm, 74–89g; ♀ 230–245mm, 90–101mm, 81–105g

A pale-colored chipmunk with a grayish head, rump and thighs. Back has contrasting pale and brown stripes. The center pair of pale stripes are grayish, the outer pair are creamy white. Head is somewhat flattened. Colors are brightest in spring, dullest in late winter. The subspecies from the Kingston mountains is smaller and darker. Redder than the Least Chipmunk. Has paler facial stripes and less contrasting dorsal stripes than Palmer's, Lodgepole, Yellow-pine, and Uinta Chipmunks. Genital bones are shown for certain ID. Call rate lower (eight syllables) than Palmer's Chipmunk. Lives in pinyon pine and juniper habitat at high elevations (1500–2600m).

genital bones

♂

♀

MERRIAM'S
CHIPMUNK

genital bones

♂

♀

CALIFORNIA
CHIPMUNK

genital
bones

♂

♀

LODGEPOLE
CHIPMUNK

winter

PALMER'S CHIPMUNK

genital bones

♂

♀

summer

PANAMINT
CHIPMUNK

genital bones

♂

♀

PLATE 29
NORTHWESTERN COASTAL CHIPMUNKS

CONFUSING CALIFORNIA CHIPMUNKS – These species are difficult to tell apart. Four (*T. ochrogenys, senex, siskiyou,* and *townsendii*) were once lumped together in the *Tamias townsendii* species group. Geographic range and color patterns are useful, but genital bones (from the skeleton) and voice are the only 100% reliable methods to distinguish between some of these overlapping species pairs. To aid in comparisons, we have placed the duller coastal morphs and the brighter inland morphs on different pages.

TOWNSEND'S CHIPMUNK *Tamias townsendii* 230–280mm, 95–120mm, 90–118g

The only chipmunk on the coast in its range, has indistinct dorsal stripes and brownish cheeks. The coastal form is dark brownish or olivaceous in color, with indistinct pale and dark lines on the back. Winter animals darker with buff-white stripes while summer animals are brighter. See next plate for details on the brighter inland form. Call has two to three syllables per chip. Lives in forests and recently cleared areas.

YELLOW-CHEEKED CHIPMUNK *Tamias ochrogenys* 233–297mm, 97–130mm, 60–118g

Has a less bushy tail, a darker back, and more distinct dorsal stripes than other *townsendii*-group chipmunks. The center stripe on back is the largest. Tail is dark rufous to bright orange ventrally and the same color as the rest of the body dorsally. Winter pelage is long, silky, dense and more dull-colored with dark tawny olive to blue-gray tones. Vocalizes with a unique sounding low-frequency two-syllable "chip." Thrives in dark, moist redwood forests.

SONOMA CHIPMUNK *Tamias sonomae* 220–264mm, 100–126mm, 63–77g

A brightly colored chipmunk with a bushy tail that is slightly edged in buff. The outer pair of pale dorsal stripes are dull whitish washed with cinnamon-buff molting to grayish in the winter; the dark dorsal stripes are black (rarely fuscous). Dorsally, tail is blackish mixed with tawny. Ventrally, tail is tawny, bordered with black and tipped with a small amount of pale buff. Call a low, labored, blunt "pok" sound. Lives in chaparral, small brushy clearings in forests, and streamside thickets. (See page 74 for the inland forms and genital bones.)

SHADOW CHIPMUNK *Tamias senex* 223–281mm, 94–122mm, 90–120g

A dark olive chipmunk with one pair of pale dorsal stripes brighter than the other in the summer. Tail is relatively thin. The dark dorsal stripes scarcely reach the rump; the central pale stripes are cinnamon while the outer pair are pale yellowish tawny. Inland populations are brighter-colored (see page 74). Calls with a rapid series of three to four syllables, sometimes stretching to 10 syllables. Primarily arboreal in humid coastal redwood forests (dull morph), and more arid inland forests (brighter morph, see page 74).

MERRIAM'S CHIPMUNK *Tamias merriami*

Large chipmunk with a bushy tail. Distinguished from Townsend's Chipmunk by having gray (not brown) cheeks. Edging on tail is buff-colored, not white. Brighter inland (see page 70 for more details). Calls with "chip"s from elevated perches in pine and oak forests.

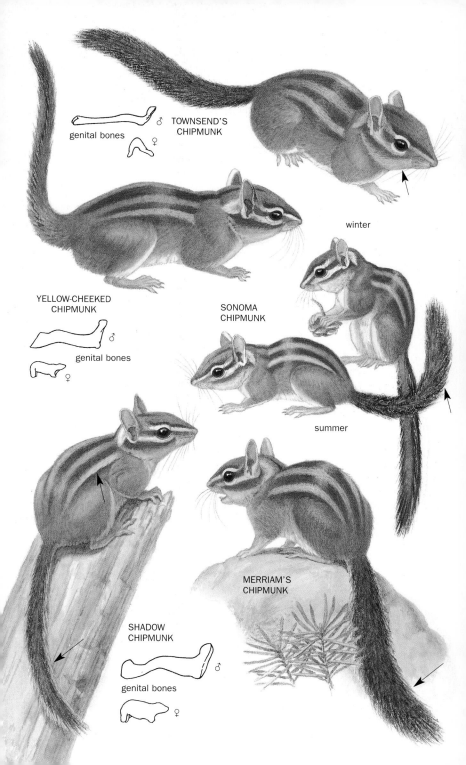

genital bones ♂
TOWNSEND'S
CHIPMUNK
♀

winter

YELLOW-CHEEKED
CHIPMUNK

SONOMA
CHIPMUNK

genital bones ♂
♀

summer

MERRIAM'S
CHIPMUNK

SHADOW
CHIPMUNK

genital bones ♂
♀

PLATE 30
NORTHWESTERN INLAND CHIPMUNKS

TOWNSEND'S CHIPMUNK *Tamias townsendii*

Larger than the overlapping inland chipmunks (*T. amoenus* and *minimus*). Paler than the coastal *townsendii* subspecies, with more reddish or orangish color. The pale body stripes fade out both near the head and rump, becoming more or less obscured by tawny or grayish shades. (*See also* the coastal chipmunks, page 72.) Lives in coniferous forest and brush.

SHADOW CHIPMUNK *Tamias senex*

Orangish, with a gray rump and bright white outer dorsal stripe. The sides are clay color to tawny; rump and thighs are gray; underparts are creamy white. Well-defined dark body stripes are brownish black; middle stripe is darkest. The middle pair of pale, back stripes are grayish white to cinnamon; outer pair are brighter, but fade to duller toward the rump. More details on the previous plate. Uses humid coastal forests (dark morph, see page 72) and arid inland forests (brighter morph).

SONOMA CHIPMUNK *Tamias sonomae*

A brightly colored chipmunk with a pale-edged tail. Back is cinnamon-buff with blackish dorsal stripes. The medial pale stripes are smoke gray and the outer pair is creamy white. The rump and thighs are gray mixed with cinnamon-buff. Underparts are creamy white. Dorsally, the tail is black, sprinkled with pale buff. Ventrally the tail is tawny and is bordered with fuscous black and edged with buff in the summer and grayish white in the winter. Lives in chaparral, small brushy clearings in forests, and streamside thickets. (*See page 72 for more details.*)

ALPINE CHIPMUNK *Tamias alpinus* 166–184mm, 63–81mm, 27–45g

Our smallest chipmunk has a short, broad tail that is frosted above and has black at the tip. Tail is shorter and flatter than *T. minimus* with larger ears and paler dorsal coloration. Smaller than *T. amoenus*, with duller coloration, more grayish hind feet, and tail with more black at tip. Winter animals are grayer above, with less tawny sides. Lives in rock-bordered alpine meadows and on talus slopes near timberline of the highest peaks of the Sierra Nevada.

LONG-EARED CHIPMUNK *Tamias quadrimaculatus*
♂ 230–239mm, 85–100mm, 74–89g; ♀ 230–345mm, 90–101mm, 81–105g

A reddish chipmunk with distinct stripes and large, slightly tufted ears. On face, pale and dark stripes contrast strongly, and the pale malar stripe connects with the large creamy-white ear patches. The top of the tail is blackish brown overlaid with grayish white; the undertail is tawny brown, bordered with mixed dark and pale gray fur. Winter animals have a grayish-brown top of the head, rump, and thighs. Summer pelage has less gray and more cinnamon in the upperparts. Call a sharp "whssst" or "pssst". Forages on ground or in bushes of chaparral and forests.

SISKIYOU CHIPMUNK *Tamias siskiyou* 250–268mm, 98–117mm, 65–85g

A large, dull-colored chipmunk with a small range. Outer lateral stripes on dorsum are grayish or brownish and paler than the inner stripes. Yellow-pine Chipmunk, the only sympatric chipmunk, is much smaller with lateral dorsal stripes that are white. Vocalizes with an intense one-syllable call that starts at a low pitch, rises, and then falls again. Uses moist coniferous forests and second-growth forests usually dominated by Douglas fir.

TOWNSEND'S CHIPMUNK

SHADOW
CHIPMUNK

SONOMA CHIPMUNK

♂

genital bones

♀

winter

summer

ALPINE
CHIPMUNK

♂
genital bones
♀

winter

summer

LONG-EARED CHIPMUNK

♂
genital bones
♀

♂ genital bones
♀

SISKIYOU CHIPMUNK

PLATE 31
NORTHERN *THOMOMYS*

GOPHERS – These small rodents are well adapted to life underground with their reduced eyes and ears and enlarged front feet and claws. They also dig with their incisors, and can close their lips around these front teeth, allowing them to continue chewing a new hole without getting a mouthful of dirt. Burrow systems are conspicuous, while the animals themselves are rarely seen. They can transport food in their fur-lined cheek pouches. The three North American genera are distinct by having 0 (*Thomomys*), 1 (*Pappogeomys*) or 2 (*Geomys*) grooves in their upper incisors. All of the *Thomomys* on this plate have skulls lacking a sphenoidal fissure (see page 79).

CAMAS POCKET GOPHER *Thomomys bulbivorus* 220–328mm, 56–99mm, 375–543g

Largest of the *Thomomys*, with a very restricted range. Dull sooty brown on the back, and dark grayish brown below, often with an irregular-shaped white patch on the throat. The ears and nose are blackish. Tips of the upper incisors angle distinctly forward in the front of the mouth, rather than pointing downward, giving the animal a bucktoothed appearance. Restricted to the Willamette Valley, and nearby tributaries in Oregon.

NORTHERN POCKET GOPHER *Thomomys talpoides* 165–260mm, 40–75mm, 60–160g

A widespread and variable gopher with dark patches behind the ear. Color can be brown or grayish brown to yellow-brown. A small *Thomomys*, with short soft fur that is less glossy than in most other species. Often has white markings under chin. Lack of a sphenoidal fissure is a feature in the skull, which separates it from other potentially overlapping gophers (see skull illustrations on page 79). This species is associated with "Mima Mounds," mounded areas up to 2m high and 20m in diameter in grassland areas. The larger the mounds, the more gophers using an area. Uses a variety of open habitat types, even forests with widely spaced trees.

WYOMING POCKET GOPHER *Thomomys clusius* 161–184mm, 50–70mm, 46–88g

Small gopher with pale yellow color, lack of dark-colored patches behind and below ears, and presence of a fringe of white hair on ears. Often associated with loose gravelly soil and greasewood habitats. Extremely limited distribution (red area on map), where it is the smallest and palest gopher.

IDAHO POCKET GOPHER *Thomomys idahoensis* 167–203mm, 40–70mm, 46–88g

Very small gopher that is pale yellowish in Idaho and Montana, but darker brown in Wyoming (see yellow area on map of Wyoming Pocket Gopher). Reddish individuals are rare. Differs from others in its range in lacking dark ear patches and contrasting grayish cheeks. Lacks the white ear fringe of the neighboring Wyoming Pocket Gopher. Seems to favor shallow and stony soils.

WESTERN POCKET GOPHER *Thomomys mazama* 191–233mm, 53–78mm, 75–125g

Richly colored gopher with pale to dark reddish-brown back and gray underparts tipped with buff. Some individuals on dark soils are darker colored. Has long ears for a gopher. Below the ear is a black patch of fur five to six times the area of the ear. Uses deep humic volcanic soils of alpine meadows and small glacial prairies.

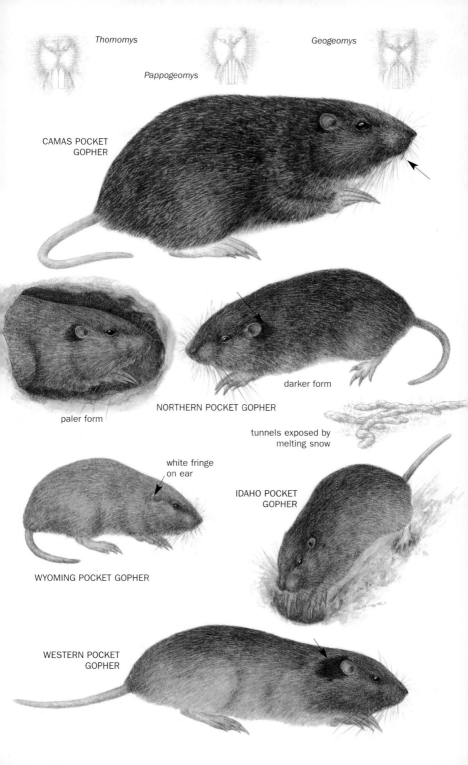

Thomomys

Pappogeomys

Geogeomys

CAMAS POCKET
GOPHER

paler form

darker form

NORTHERN POCKET GOPHER

tunnels exposed by
melting snow

white fringe
on ear

IDAHO POCKET
GOPHER

WYOMING POCKET GOPHER

WESTERN POCKET
GOPHER

PLATE 32
PAPPOGEOMYS AND
SOUTHERN *THOMOMYS*

BOTTA'S POCKET GOPHER *Thomomys bottae* ♂ 170–280mm, 62–92mm, 110–250g; ♀ 150–240mm, 55–73mm, 80–160g

One of the world's most variable small mammals, usually has a unique purple tinge to the flanks. Ranges from dark blackish brown through various shades of reddish and yellowish browns, pale grays, and yellows, to nearly white. Ventral coloration usually mirrors that of back. Patches of white on throat, chest, or abdomen are not uncommon. Body color is tied to soil color. Dark patches behind ear are most evident in lightly colored individuals. Skull has an obvious sphenoidal fissure. Typically larger than the Southern Pocket Gopher and smaller than Townsend's Pocket Gopher. Lives in virtually any friable soil, from rich valleys to rocky montane slopes and desert outwash areas. Wide range of habitat types from desert up to coniferous forests, but mainly in open areas where soils are deep enough for burrows (meadows, stream sides, etc.).

SOUTHERN POCKET GOPHER *Thomomys umbrinus*
♂ 210–250mm, 65–80mm, 110–175g; ♀ 180–230mm, 55–70mm, 80–120g

Typically has a unique darkened, bluish mid-dorsal band. Color varies from dark, rich reddish brown to blackish brown. Skull has a sphenoidal fissure. Only has three pairs of pectoral mammae, as opposed to four pairs normally found in Botta's Pocket Gopher. Known in United States only from Pajarito, Santa Rita, Patagonia, and Huachuca mountains in Arizona, and Animas Mountains in New Mexico. Inhabits desert scrub and grasslands through intermediate elevation oak and pine-oak woodlands, to grassy meadows in high elevation pine and fir forests.

TOWNSEND'S POCKET GOPHER *Thomomys townsendii*
♂ 232–315mm, 60–99mm, 201–417g; ♀ 226–287mm, 58–92mm, 122–308g

Larger than other Great Basin pocket gophers, nose is usually darkened. Color varies from pale grayish tan through cinnamon to very dark brown. In some areas a white blaze on the top of the head is fairly common. The nose is sooty black, gray, or dark brown, and most animals also have a black spot behind each ear. Typically, the tops of the feet and tail are white and there is a white chin patch. Skull has a sphenoidal fissure. Uses deep, moist soils of river valleys and prehistoric lakebeds.

MOUNTAIN POCKET GOPHER *Thomomys monticola* 190–227mm, 55–95mm, 75–105g

A uniform brown gopher with relatively large and pointed (not rounded) ears. Behind the ears there is a large black patch about three times the size of the ear. There is no sphenoidal fissure in the skull. The snout is darker than the face. Favors montane meadows, pastures, and rocky slopes of pine, fir, and spruce in the Sierra Nevadas.

YELLOW-FACED POCKET GOPHER *Cratogeomys castanops*
220–315mm, 60–95mm, ♂ 385–410g; ♀ 225–290g

Large eyes and a single groove in the upper incisors distinguish this species from all other pocket gophers. The upperparts vary from pale yellowish buff to dark reddish brown, and the underparts are whitish to bright orange. All hairs are grayish at the base, usually with slightly darker hues on the back. It uses deep sandy soil with few rocks, in both grasslands and more arid habitats.

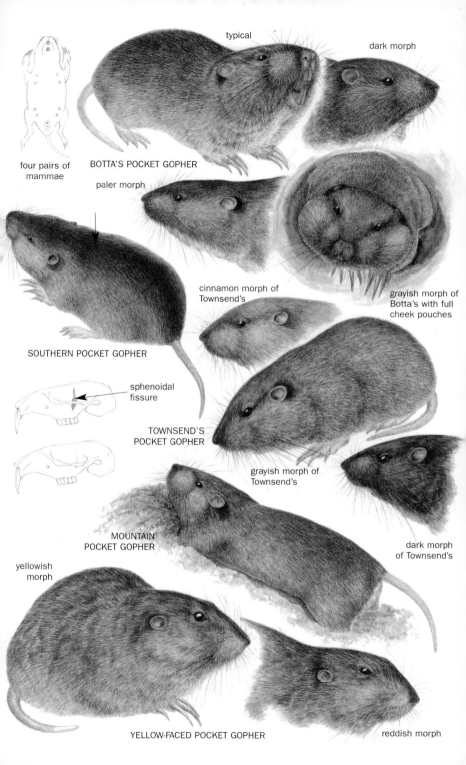

typical

dark morph

four pairs of
mammae

BOTTA'S POCKET GOPHER

paler morph

SOUTHERN POCKET GOPHER

cinnamon morph of
Townsend's

grayish morph of
Botta's with full
cheek pouches

sphenoidal
fissure

TOWNSEND'S
POCKET GOPHER

grayish morph of
Townsend's

MOUNTAIN
POCKET GOPHER

dark morph
of Townsend's

yellowish
morph

YELLOW-FACED POCKET GOPHER

reddish morph

PLATE 33
GEOMYS

TRICKY TEXAS GOPHERS – The first five species of long-tailed gophers are virtually indistinguishable by appearance, and were originally identified by genetic differences. They can be distinguished from other overlapping gophers by size and by skull characters (all lack sagittal crest and knob on zygoma).

BAIRD'S POCKET GOPHER *Geomys breviceps* 192–222mm, 54–67mm, 78–150g

Limited to sandy soils, or sandy loams in prairie grasslands and oak savannas (yellow area on map).

ATTWATER'S POCKET GOPHER *Geomys attwateri* 192–235mm, 51–70mm, 102–170g

Limited range in Texas (see red area on map of Baird's Pocket Gopher), where it occupies friable soils in grasslands.

KNOX JONES'S POCKET GOPHER *Geomys knoxjonesi* 203–282mm, 57–104mm, 160–185g

Restricted to areas of western Texas and eastern New Mexico with deep, sandy soils (yellow area on map). Favors yucca grasslands, although can now be found along grassy roadsides, and in lawns and pastures.

CENTRAL TEXAS POCKET GOPHER *Geomys texensis* 185–272mm, 51–80mm, 125–165g

Smaller than nearby Attwater's, Plains, and Texas Pocket Gophers. Favors deep brown loamy sands or gravelly sandy loams in live-oak, mesquite, ash, and juniper habitats as well as blackbrush and desert hackberry communities (see red area on map of Knox Jones's).

PLAINS POCKET GOPHER *Geomys bursarius* 225–325mm, 60–121mm, 120–250g

Sparsely haired tail is about one third of body length. Inhabits deep sandy or loamy soils in most Great Plains habitats, as well as roadsides, lawns, and pastures.

TEXAS POCKET GOPHER *Geomys personatus* 216–360mm, 62–125mm, 156–400g

Medium-sized, drab brown gopher with soft, short pelage that is pale to whitish on the underparts. Very similar to Attwater's, but skull has a distinct sagittal crest, and it differs in habitat, occupying only deep, sandy soils in coastal and river bottoms in southern Texas (red area on map).

DESERT POCKET GOPHER *Geomys arenarius* 218–302mm, 52–106mm, 165–254g

Distinguished from the other pocket gophers in its range by having two grooves on the upper incisors. Inhabits sandy and disturbed soils along the upper Rio Grande Valley (see yellow area on map of Texas), where it occupies desert scrub as well as a variety of man-made habitats.

SOUTHEASTERN POCKET GOPHER *Geomys pinetis* 215–324mm, 57–120mm, 135–208g

Only gopher in range. Inhabits deep, sandy soils in long-leaf pine forests. Can be a pest in orchards and lawns.

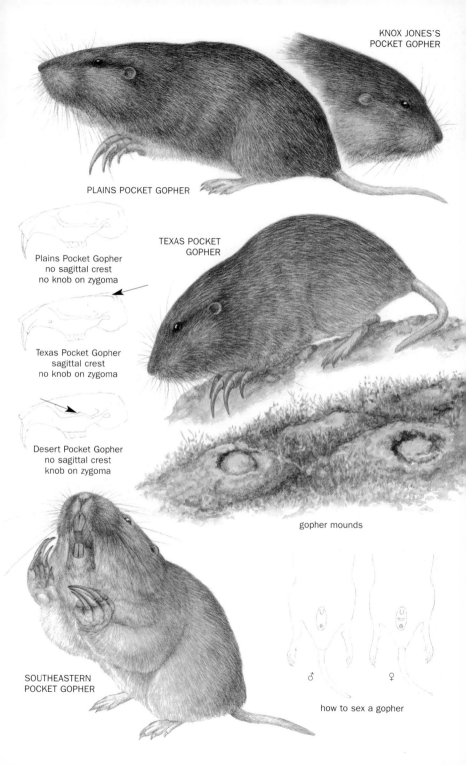

KNOX JONES'S POCKET GOPHER

PLAINS POCKET GOPHER

TEXAS POCKET GOPHER

Plains Pocket Gopher
no sagittal crest
no knob on zygoma

Texas Pocket Gopher
sagittal crest
no knob on zygoma

Desert Pocket Gopher
no sagittal crest
knob on zygoma

gopher mounds

SOUTHEASTERN POCKET GOPHER

♂   ♀

how to sex a gopher

PLATE 34
WESTERN DESERT *PEROGNATHUS*

POCKET MICE – These small, nondescript mice are quite diverse, and sometimes hard to tell apart. All have relatively long feet, but are poor jumpers compared to their cousins the kangaroo rats. Pocket mice are typically distinguished by the amount of spiny guard hairs present (i.e., spiny or smooth) and by subtle differences in the markings on their bodies and tails. They use their fur-lined cheek pouches (also known as pockets) to carry seeds back to underground larders.

WHITE-EARED POCKET MOUSE *Perognathus alticolus* 130–183mm, 70–97mm, 16–24g

Rarest of the *Perognathus*, this species resembles Great Basin Pocket Mouse, but is restricted to isolated mountain ranges bordering the Mojave Desert. With white or yellowish hair on the ears, it also differs from San Joaquin and Little Pocket Mouse in having a lobed antitragus and a somewhat longer, darker crest on the tail. Known from open grassland and upland arid shrub communities between 1000 and 2000m.

SAN JOAQUIN POCKET MOUSE *Perognathus inornatus* 128–160mm, 63–78mm, 7–12g

Medium-sized mouse whose tail averages longer than head+body. Pelage is soft, with upperparts yellowish to pink overlaid with blackish hairs; the extent of the overlay determines the overall tone in various subspecies. Lateral line is moderately well marked, and underparts are white, with tail faintly bicolored. Posterior third of sole of hind foot is haired, and whiskers are rather short. Uses arid annual grassland, savanna, and desert scrub, with sandy washes, fine soils, and scattered vegetation.

ARIZONA POCKET MOUSE *Perognathus amplus* 135–173mm, 75–88mm, 9–14g

Small pocket mouse with orangish-tan upperparts sprinkled with black to varying degrees, and white or pale tan underparts. Tail is longer than head and body, slightly bicolored, and lacks a terminal tuft. Forages between small shrubs or bunch-grasses in flat habitats with fine-textured soils, venturing into open areas only when the moon is dark.

LITTLE POCKET MOUSE *Perognathus longimembris* 110–151mm, 56–86mm, 6–11g

Smallest pocket mouse in its range, with coloration varying from gray to reddish-brown to cream dorsally, with paler buff or white hairs on undersurface. Tail relatively longer than that of San Joaquin Pocket Mouse, hind foot averages shorter than that of Arizona Pocket Mouse. The subspecies that uses sandy habitats in southern California (*P. l. pacificus*) was once thought extinct, but a few endangered populations were rediscovered in 1993. Uses open grassland, shrub-steppe, and coastal sage habitats, in addition to very arid desert areas.

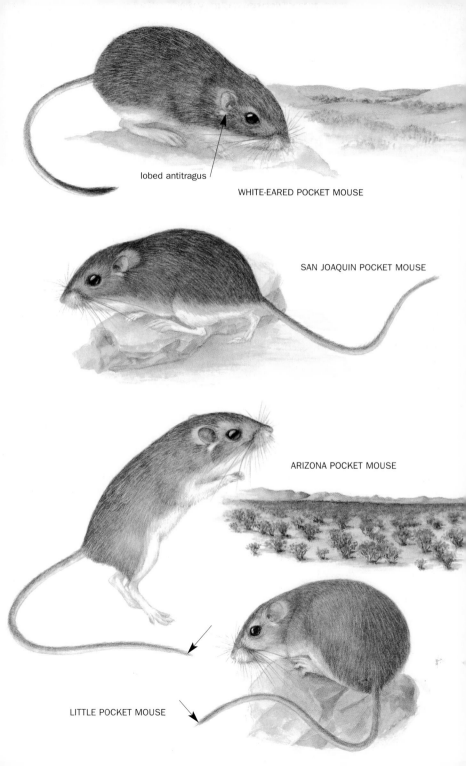

lobed antitragus

WHITE-EARED POCKET MOUSE

SAN JOAQUIN POCKET MOUSE

ARIZONA POCKET MOUSE

LITTLE POCKET MOUSE

PLATE 35
GREAT PLAINS *PEROGNATHUS*

OLIVE-BACKED POCKET MOUSE *Perognathus fasciatus*
125–142mm, 57–68mm, 8–14g

Resembles other *Perognathus* in having soft pelage, with no spines or bristles, and somewhat hairy soles of feet. The olive-colored back distinguishes it from the otherwise similar Plains Pocket Mouse, and a yellowish lateral stripe separates the pure white belly. Resident of grassland, shrub-steppe, and desert scrub habitats of northern Great Plains and intermontane west.

PLAINS POCKET MOUSE *Perognathus flavescens* 117–155mm, 50–89mm, 7–16g

Small, soft-furred pocket mouse that is somewhat larger than Silky and Merriam's Pocket Mouse, with a relatively longer tail, and smaller buffy patches behind the ears. Occupies sand dunes and other stabilized, sandy soils in the Great Plains and mountain states. Found from grassland and desert scrub up through oak woodlands and into pinyon-juniper communities.

MERRIAM'S POCKET MOUSE *Perognathus merriami* 95–121mm, 42–61mm, 5–9g

Tiny pocket mouse with yellowish-orange fur tinged with black on the back, and underparts that are white or pale buff and clearly separated from the darker sides. Very similar to Silky Pocket Mouse, but with a relatively longer tail, shorter, slightly coarser pelage that is paler and more yellowish, and slightly smaller buffy spots behind the ears. Common in short-grass prairies, desert scrub, and open, arid, brushy areas.

SILKY POCKET MOUSE *Perognathus flavus* 100–130mm, 44–60mm, 5–10g

This is one of the smallest pocket mice, and among the smallest rodents in North America. Looks essentially identical to Merriam's Pocket Mouse. Yellowish or reddish brown on back and sides, and white on the belly, with buffy patches behind each ear. Occupies grassy and shrubby habitats in western and southern plains and southwestern intermontane areas.

GREAT BASIN POCKET MOUSE *Perognathus parvus*
♂ 160–181mm, 85–97mm, 21–31g; ♀ 160–190mm, 85–90mm, 16–29g

This is the largest *Perognathus*. Differs from others in having hind feet longer than 20mm, buffy hair on the inside of the ears, a lobed antitragus, an olivaceous lateral line and a bicolored tail that is is dark above and has only a slight terminal tuft. Occupies arid and semiarid sandy areas of sagebrush, steppe, open shrub, woodland, deserts, and dry grasslands.

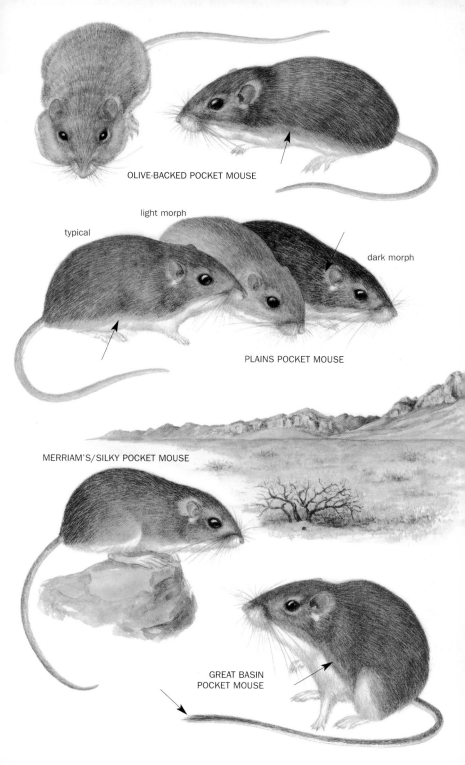

OLIVE-BACKED POCKET MOUSE

light morph

typical

dark morph

PLAINS POCKET MOUSE

MERRIAM'S/SILKY POCKET MOUSE

GREAT BASIN
POCKET MOUSE

PLATE 36
LIOMYS AND *SPINY* *CHAETODIPUS*
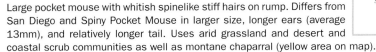

CALIFORNIA POCKET MOUSE *Chaetodipus californicus*
190–235mm, 103–143mm, 18–29g

Large pocket mouse with whitish spinelike stiff hairs on rump. Differs from San Diego and Spiny Pocket Mouse in larger size, longer ears (average 13mm), and relatively longer tail. Uses arid grassland and desert and coastal scrub communities as well as montane chaparral (yellow area on map).

SAN DIEGO POCKET MOUSE *Chaetodipus fallax* 176–200mm, 88–118mm, 17–22g

Medium-sized pocket mouse with broad region of yellowish to orangish hair on its sides that contrasts with its dark brown back. Has a number of stiff bristly hairs or spines in the rump region, but fewer than Spiny Pocket Mouse. Uses sparse, low, desert shrublands up to dense, high, coastal, sage-scrub vegetation (see red area on map of California Pocket Mouse).

NELSON'S POCKET MOUSE *Chaetodipus nelsoni* 182–193mm, 104–117mm, 14–18g

Drab gray, medium-sized mouse with harsh pelage and a distinctly tufted tail that is longer than the head and body. Soles of hind feet are black. Numerous and prominent spines on the rump; distal ends of the rump spines usually dark-colored dorsally; entire rump spine is pale-colored laterally. White spots below the ears. Uses rocky places in Chihuahuan desert shrub vegetation with rocky soils and cactus, creosote, sotol, and lechuguilla provide scattered cover. Avoids sandy soils.

SPINY POCKET MOUSE *Chaetodipus spinatus* 164–225mm, 89–128mm, 13–18g

Upper pelage is drab brown and shaggy. Hairs are dark gray near base, pale tan in middle, black at tips. Lateral line is faint or absent. Underparts are buff-white. Ears are small (average 10mm) and dusky, and there is a small white spot at the base of each ear. Tail is bicolored, with a distinct crest near the tip. Spines are located mostly on the rump, but scattered spines occur as far forward as the shoulder region. Inhabits rough desert landscapes of boulders, washes, rocky slopes, coarse soil, and sparse vegetation.

ROCK POCKET MOUSE *Chaetodipus intermedius* 157–188mm, 84–112mm, 10–20g

Medium-sized mouse with drab grayish-brown fur on back, a pale orange-brown line on the sides, and white underneath. Comparatively harsh fur with weak "spines" on the rump, and soles of hind feet are naked to the heels. Tail is longer than the head and body and distinctly tufted at the tip. Uses rocky gulches, canyons, or boulders and rarely found on sandy or silty soils.

MEXICAN SPINY POCKET MOUSE *Liomys irroratus*
♂ 216–262mm, 106–138mm, 40–60g; ♀ 207–251mm, 102–131mm, 35–50g

Liomys is easily distinguished from all other pocket mice by ungrooved upper incisors. Grayish brown with white underparts, separated by buff stripe between the darker upperparts and paler underside. Fur on back has a harsh appearance caused by the mix of stiff spiny hairs and soft slender hairs. Unique spoon-shaped claw on hind foot. Uses dense brushy areas along old river terraces, or in subtropical palm forests in extreme south Texas.

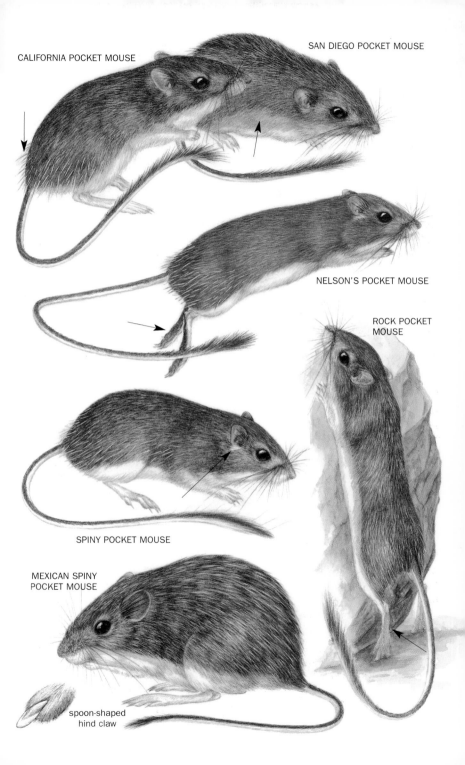

CALIFORNIA POCKET MOUSE

SAN DIEGO POCKET MOUSE

NELSON'S POCKET MOUSE

ROCK POCKET MOUSE

SPINY POCKET MOUSE

MEXICAN SPINY POCKET MOUSE

spoon-shaped hind claw

PLATE 37
SMOOTH CHAETODIPUS

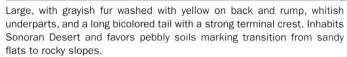

BAILEY'S POCKET MOUSE *Chaetodipus baileyi* ♂ 206–240mm, 76–140mm, 25–38g; ♀ 176–228mm, 86–125mm, 24–37g

Large, with grayish fur washed with yellow on back and rump, whitish underparts, and a long bicolored tail with a strong terminal crest. Inhabits Sonoran Desert and favors pebbly soils marking transition from sandy flats to rocky slopes.

LONG-TAILED POCKET MOUSE *Chaetodipus formosus*
172–211mm, 86–125mm, 17–25g

Medium-sized mouse with soft pelage and no stiff bristly hairs on rump. Tail is long, with a distal crest and conspicuous terminal tuft. Smaller than Bailey's Pocket Mouse and with longer ears than Desert Pocket Mouse. Uses dry and rocky areas, such as lava beds, desert scrub, dry stream beds, and boulder-strewn hillsides in Great Basin and Mojave and Colorado deserts.

HISPID POCKET MOUSE *Chaetodipus hispidus* 198–223mm, 90–113mm, 30–47g

Large pocket mouse, with distinctly coarse, but not spiny, pelage. The tail is only slightly shorter than the body and distinctively bicolored with no terminal tuft. Upperparts are olive-buff and separated from white underparts by a distinct orangish-yellow stripe. Occupies grassland habitats from desert areas up through pinyon-juniper zones.

CHIHUAHAN POCKET MOUSE *Chaetodipus eremicus* 170–215mm, 90–115mm, 15–23g

Slightly larger and lighter in color than the very similar Desert Pocket Mouse; best distinguished by geographic range. The two were only recently recognized as distinct species from genetic evidence.

DESERT POCKET MOUSE *Chaetodipus penicillatus* 155–185mm, 83–110mm, 13–20g

Medium-sized pocket mouse lacking spines on rump, with yellowish-brown to yellowish-gray upper pelage and whitish underparts. Tail is long, bicolored, and strongly crested. Uses sandy soils with creosote, mesquite, or palo verde vegetation, especially along desert washes. Rocky soils tend to be avoided.

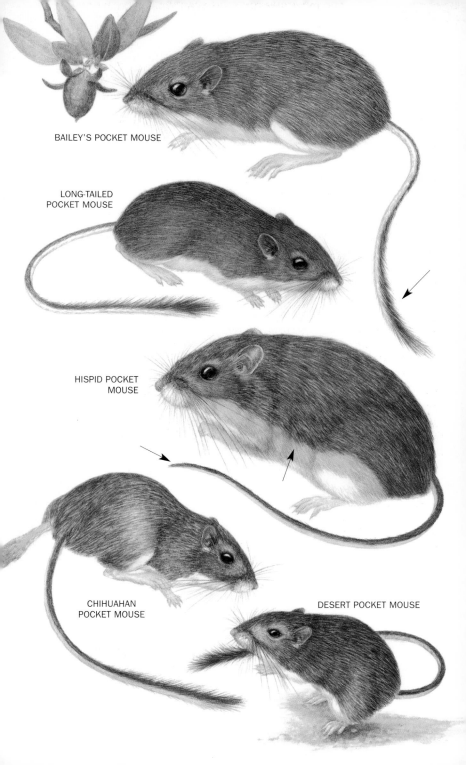

BAILEY'S POCKET MOUSE

LONG-TAILED
POCKET MOUSE

HISPID POCKET
MOUSE

CHIHUAHAN
POCKET MOUSE

DESERT POCKET MOUSE

PLATE 38
KANGAROO MICE AND SMALL FOUR-TOED KANGAROO RATS

KANGAROO MICE – With their huge hind feet and long tails, kangaroo mice are miniature versions of five-toed kangaroo rats. They differ by size, and by having a fat deposit in the tail that makes it appear wider in the middle, and with the soles of the hind feet well furred. The two species of kangaroo mice are distinguished mainly by color. They are nocturnal seed eaters, and can transport food in their fur-lined cheek pouches.

DARK KANGAROO MOUSE *Microdipodops megacephalus*
138–177mm, 67–103mm, 10–17g

A dark-colored kangaroo mouse. Brownish, blackish, or grayish dorsal pelage distinguish it from the Pale Kangaroo Mouse. Also, the dorsal surface of the tail is darker than the body, and has a black tip. The hind foot is less than 25mm. Hair on the underparts is dark at the base and white-tipped. Uses Upper Sonoran sagebrush desert, on fine gravelly soils.

PALE KANGAROO MOUSE *Microdipodops pallidus* 150–173mm, 74–99mm, 10–17g

A pale pinkish-cinnamon-colored kangaroo mouse. Dorsal surface of tail is same color as the body, and lacks a dark tip. Hind foot is more than 25mm. Belly fur is pale pinkish cinnamon at the base. Uses valley bottomlands with stabilized dunes with fine sand in Upper Sonoran life zone dominated by saltbush and greasewood.

KANGAROO RATS – Aptly named rats with enormous hind feet, miniature front feet, and long tails. They have large eyes, a white racing stripe along their flanks and tails, and clearly marked dermal glands on neck between shoulder blades. Species are typically identified by their size, the number of toes on the hind feet (four or five), and characteristics of their tail. All spend the day in burrows and the night foraging for seeds, which may be transported in their fur-lined cheek pouches. May leap up to 2m in a single bound.

MERRIAM'S KANGAROO RAT *Dipodomys merriami* 195–282mm, 120–182mm, 33–54g

A relatively small, four-toed, slender-footed, and usually buff-colored kangaroo rat. The tail is much longer than head and body, ending with a dusky to dark tuft at the tip. Body has dorsal and ventral dusky stripes, and the underparts are white. Smaller than most four-toed kangaroo rats, with a dark (not white)-tipped tail. Different from the San Joaquin Valley Kangaroo Rat by range, body size, and color pattern on the tail. The subspecies from the San Bernardino Mountains (*D. m. parvus*) is Endangered. Occupies a wide variety of soil types and habitats in the Southwest.

SAN JOAQUIN VALLEY KANGAROO RAT *Dipodomys nitratoides*
211–253mm, 120–152mm, 40–53g

A small, four-toed kangaroo rat with yellowish-brown upperparts and white underparts. Facial crescents are dusky and meet over the bridge of the nose. Upper and lower tail stripes are sooty blackish, meeting along the terminal third, thus interrupting the white side stripes. Inner aspect of the hind legs to heel are dull fulvous, and the underparts of the thigh stripes are white. Uses alkaline plains sparsely covered with grass, or saltbush and other arid vegetation.

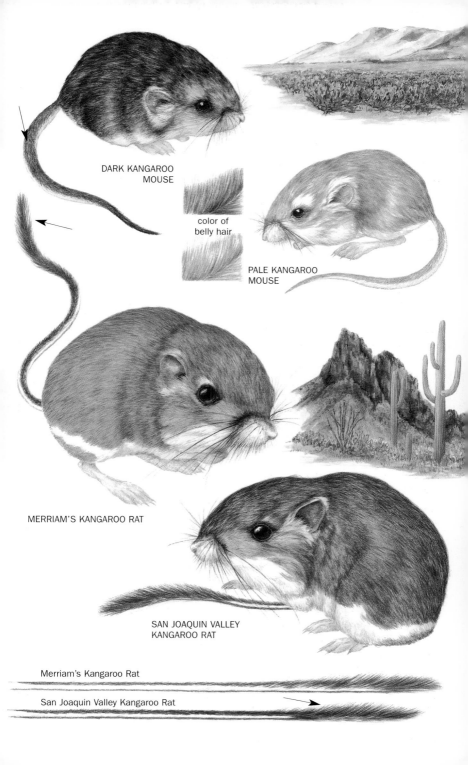

DARK KANGAROO
MOUSE

color of
belly hair

PALE KANGAROO
MOUSE

MERRIAM'S KANGAROO RAT

SAN JOAQUIN VALLEY
KANGAROO RAT

Merriam's Kangaroo Rat

San Joaquin Valley Kangaroo Rat

PLATE 39
LARGE FOUR-TOED KANGAROO RATS

BANNER-TAILED KANGAROO RAT *Dipodomys spectabilis*
♂ 315–349mm, 185–208mm, 110–132g; ♀ 310–345mm, 180–205mm, 98–130g

A large, spectacular four-toed kangaroo rat with a long, white-tipped tail. It is yellowish-brown above and white underneath, with a long tail covered with short hairs at the base and long hairs at the tip. Spots above the eye and behind the ear, hip stripes, forelimbs, dorsal surface and sides of hind feet, lateral tail stripes, ventral surface, and tip of the tail are pure white. Young animals are grayish on the back and slightly brighter on the sides. Lives in desert grasslands with scattered shrubs, and makes large mounds that contain extensive burrow systems.

DESERT KANGAROO RAT *Dipodomys deserti* 305–377mm, 180–215mm, 83–148g

Similar to *D. spectabilis*, but slightly paler overall and with paler, less contrasting dorsal and ventral stripes on tail. Large hind feet are covered with relatively long hairs, and have four toes. Indistinct white spot over eye, and another behind ear that extends across the shoulder to the white underparts. Indistinct white band across the hips and an indistinct darker spot at the base of the whiskers. Soles of hind feet are nearly white. Color of upper parts varies from pale fawn to grayish black, depending on subspecies. Uses loose sandy soil in the most arid areas of North America.

TEXAS KANGAROO RAT *Dipodomys elator* 260–345mm, 161–205mm, 65–90g

Tail is thick and long with dark stripes above and below, and white stripes on the side that end in a white tuft. Similar to Ord's Kangaroo Rat, but that species has five toes on the hind feet and a dark terminal tail tuft. Belly is white, back is a buff color interspersed with black. Nose and eye rings are black; white thigh patches are present and meet at the base of the tail. Now inhabits only three counties in central Texas; human-induced land change has driven them to local extinction in north-central Texas and southwestern Oklahoma. Uses clay and clay-loam soils with sparse vegetation, in what were originally mesquite grasslands.

CALIFORNIA KANGAROO RAT *Dipodomys californicus* 260–340mm, 152–217mm, 60–85g

Medium-large kangaroo rat with relatively broad face, darker upperparts and white belly. Tail has broad, dark dorsal and ventral stripes and a distinct white tuft. Typically has four toes on the hind feet, but five-toed individuals are occasionally found. Larger and darker than Merriam's Kangaroo Rat, and smaller and darker than the Desert Kangaroo Rat. Heermann's Kangaroo Rat is very similar, but has five toes. Uses chaparral and other shrub, but is restricted to places where open areas are available.

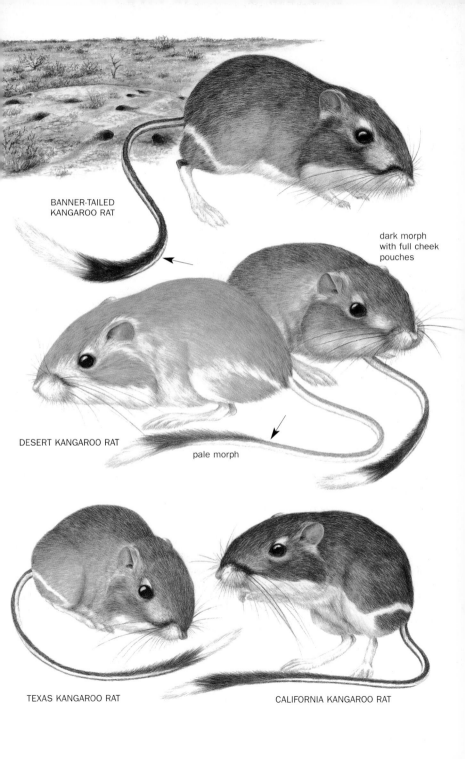

BANNER-TAILED
KANGAROO RAT

dark morph
with full cheek
pouches

DESERT KANGAROO RAT

pale morph

TEXAS KANGAROO RAT

CALIFORNIA KANGAROO RAT

PLATE 40
MID-CALIFORNIA FIVE-TOED KANGAROO RATS

HEERMANN'S KANGAROO RAT *Dipodomys heermanni*
250–313mm, 160–200mm, 70–80g

Medium-large kangaroo rat with relatively broad face and moderate-sized ears. Has five toes on the hind feet. Dorsal pelage varies from tawny-olive strongly overwashed with black to orangish yellow. Tail crest may be dark and only slightly crested, blackish and very scantily haired or white tipped. Ears are smaller (12–15mm) than those of the Big-eared and Narrow-faced Kangaroo Rats. Smaller than the Giant Kangaroo Rat and larger than the San Joaquin Valley Kangaroo Rat. The subspecies *morroensis* from the Morro Bay area is Endangered, and typically lacks a hip stripe. Occurs in wide range of habitats from Lower Sonoran to Transition zone.

GIANT KANGAROO RAT *Dipodomys ingens* 312–348mm,
157–197mm, 93–195g

Largest of the kangaroo rats. Has a dusky-colored nose, whitish cheeks and blackish eyelids. Lateral white stripes along the tail are only slightly narrower than the dorsal and ventral dark stripes. Has five toes on the hind feet. Tail tuft is dark in appearance because of a mix of pale and dark hairs. This Endangered species is restricted to southwestern San Joaquin Valley and adjacent arid areas of the Inner Coastal ranges of California. Uses sandy loamy soil on level and gently sloping ground vegetated with annual grasses and forbs and widely scattered shrubs.

BIG-EARED KANGAROO RAT *Dipodomys elephantinus*
305–336mm, 183–210mm, 79–91g

A large-bodied, big-eared, long-tailed, kangaroo rat that is moderately dark in color. Has five toes on the hind feet. Very restricted range in southern California. Upperparts are cinnamon, ears are mostly brownish, and underparts are white. Tail is heavily tufted and crested. Ventral stripe at the end of tail is narrower than the lateral white stripes. Larger, darker, and has longer ears (19mm) than Heermann's Kangaroo Rat. Paler and less conspicuously marked than the Narrow-faced Kangaroo Rat. Occurs only in southern parts of the Gabilan Mountain range in San Benito and Monterey counties, California. Uses chaparral-covered slopes, especially areas with dense vegetation.

NARROW-FACED KANGAROO RAT *Dipodomys venustus*
293–332mm, 175–203mm, 68–97g

Large, dark-colored, five-toed kangaroo rat with black nose that merges into a black band at the base of the whiskers. Top of head and back are darker, sides and thigh patches are yellowish brown. Ears are large and nearly black, with pale spots at the base and at the top of the fold. Darker than the Agile Kangaroo Rat, with bolder face markings, larger ears, longer tail, and longer rostrum. Previously considered a subspecies of the Big-eared Kangaroo Rat, but differs by being smaller and darker, with less white on cheeks, blacker ears, and bolder face markings. Darker than Heermann's Kangaroo Rat, and with much larger ears. Small geographic range along the California coast up to 1770m in elevation. Uses slopes with chaparral, oaks, or digger pine on sandy soils.

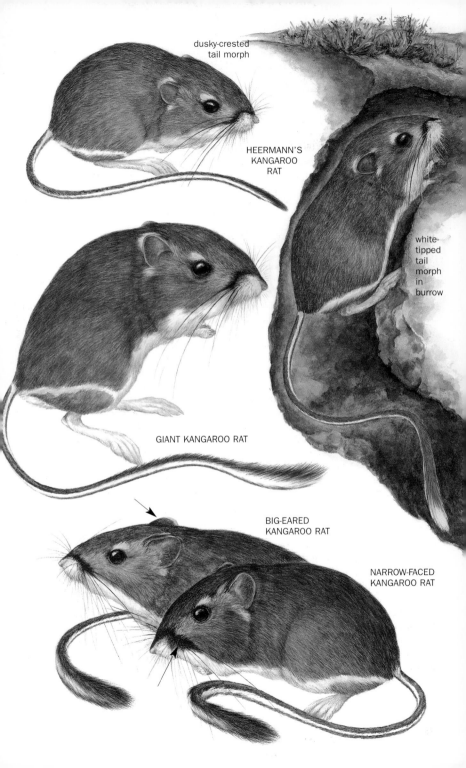

dusky-crested
tail morph

HEERMANN'S
KANGAROO
RAT

white-
tipped
tail
morph
in
burrow

GIANT KANGAROO RAT

BIG-EARED
KANGAROO RAT

NARROW-FACED
KANGAROO RAT

PLATE 41
SOUTHERN CALIFORNIA FIVE-TOED KANGAROO RATS

DULZURA KANGAROO RAT *Dipodomys simulans* 273–302mm, 160–181mm, 55–70g

Intermediate in size, with long, bicolored tail that has a blackish crest and tuft. Has five toes on the hind feet. Upperparts vary from pale grayish brown to dark reddish brown, depending on geography. Smaller than the Agile Kangaroo Rat. Longer ears (15–20mm) than Stephens's Kangaroo Rat. Extends from Baja California, Mexico, into southern California. Uses coastal chaparral and grasslands on gravelly or sandy soil.

STEPHENS'S KANGAROO RAT *Dipodomys stephensi* 277–300mm, 164–180mm, 45–73g

Medium-sized five-toed kangaroo rat with yellowish-brown upperparts overlain with black hairs. Narrow lateral white tail stripes are indistinctly demarcated from the dark stripes (not sharply demarcated broad lateral stripes as in the Agile Kangaroo Rat). Has smaller ears (12–14mm) and occupies more open habitats than the Agile and Dulzura Kangaroo Rats. Seriously Endangered by agricultural and urban development, now restricted to Riverside, San Bernardino, and San Diego counties, California. Found in sparse grasslands and coastal sage-scrub habitats.

AGILE KANGAROO RAT *Dipodomys agilis* 277–320mm, 170–195mm, 63–79g

Intermediate-sized five-toed kangaroo rat with large ears and dark reddish-brown upperparts. The belly is white, and the long, bicolored tail has a blackish crest and tuft, with broad, sharply demarcated white lateral stripes. Pelage of young animals is darker than adults, with hairs on tail not elongated into a crest. Larger than the Dulzura Kangaroo Rat, with longer ears (16–19mm) than Stephens's Kangaroo Rat. Found primarily in areas of loose soil in open chaparral and coastal sage-scrub.

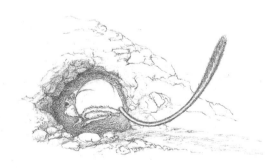

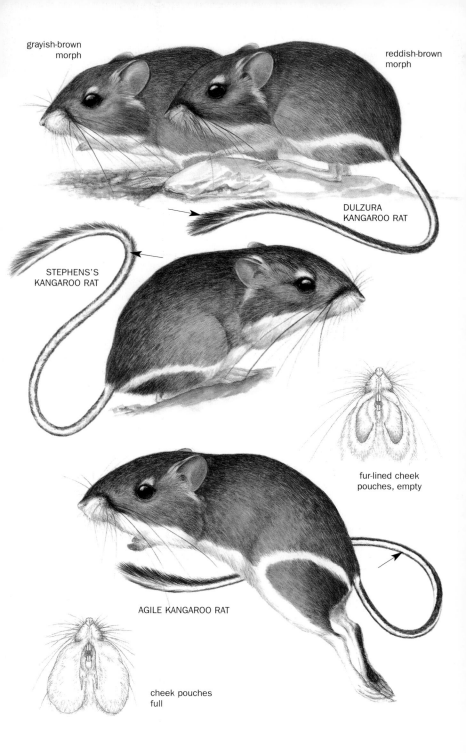

grayish-brown morph

reddish-brown morph

DULZURA KANGAROO RAT

STEPHENS'S KANGAROO RAT

fur-lined cheek pouches, empty

AGILE KANGAROO RAT

cheek pouches full

PLATE 42
OTHER FIVE-TOED KANGAROO RATS

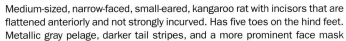

CHISEL-TOOTHED KANGAROO RAT *Dipodomys microps*
245–295mm, 135–175mm, 40–70g

Medium-sized, narrow-faced, small-eared, kangaroo rat with incisors that are flattened anteriorly and not strongly incurved. Has five toes on the hind feet. Metallic gray pelage, darker tail stripes, and a more prominent face mask distinguish it from Ord's Kangaroo Rat. Long whiskers maintain contact with the ground even when the animal is in midair. Hind feet and tail are shorter than those of Panamint Kangaroo Rat. Eats more leaves than other kangaroo rats, and is one of few mammals adapted to feed on saltbush, *Atriplex*. Occupies desert valleys dominated by saltbush in the Great Basin.

PANAMINT KANGAROO RAT *Dipodomys panamintinus*
285–334mm, 156–202mm, 57–95g

Medium-sized five-toed kangaroo rat with pale, clay-colored upperparts tinged with pale ocher. Thigh patches are large and colored like the back. Facial crescents and end of nose are broadly blackish, but are not con-tinuous (or just barely continuous) over the sides of the nose. Inner sides of legs are dusky pale brownish and underparts are white. Tail has a ventral stripe that may extend to the end of the tail vertebrae. Has a longer tail than the Gulf Coast Kangaroo Rat and longer feet (42–48mm) than Ord's Kangaroo Rat. Lower incisors are rounded and curved like most other kangaroo rats, but unlike the Chisel-toothed Kangaroo Rat. Uses coarse sand and gravelly desert flats with scattered desert scrub vegetation.

ORD'S KANGAROO RAT *Dipodomys ordii* 208–365mm, 100–163mm, 50–96g

Medium-sized, relatively short-tailed five-toed kangaroo rat. Color of back may be brownish, reddish, or blackish, depending on the subspecies. The belly is white. Has shorter hind feet (39–44mm) than the Panamint Kan-garoo Rat. The Chisel-toothed Kangaroo Rat has chisel-shaped lower incisors; the Gulf Coast Kangaroo Rat has an orange cast to the pelage and a shorter, less-crested tail. Occu-pies variety of habitats associated with fine-textured, sandy soils including semiarid grass-lands, mixed grasslands, and scrublands.

GULF COAST KANGAROO RAT *Dipodomys compactus* 205–266mm, 104–135mm, 44–60g

Medium-sized kangaroo rat with a short tail. Found only in southern Texas. Coastal island forms are orangish yellow or grayish cream on the back; main-land animals are reddish yellow on the back. This color covers the entire back and is purest on the sides and flanks. Upperparts are lightly washed with black. Ears, underside of feet, and dorsal and ventral tail stripes are similar to the dorsal color. Cheeks are white. Tail is shorter and less hairy than Ord's Kangaroo Rat. Has five toes on the hind feet. Uses sparsely vegetated areas with sandy soil, such as open mesquite savanna.

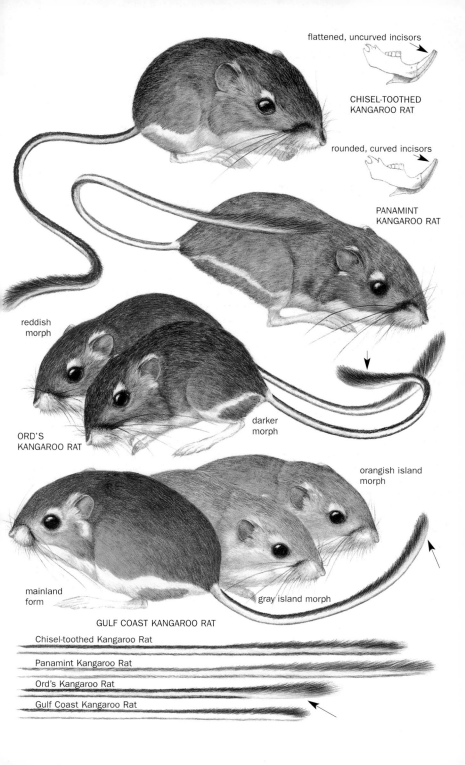

flattened, uncurved incisors

CHISEL-TOOTHED
KANGAROO RAT

rounded, curved incisors

PANAMINT
KANGAROO RAT

reddish
morph

ORD'S
KANGAROO RAT

darker
morph

orangish island
morph

mainland
form

gray island morph

GULF COAST KANGAROO RAT

Chisel-toothed Kangaroo Rat

Panamint Kangaroo Rat

Ord's Kangaroo Rat

Gulf Coast Kangaroo Rat

PLATE 43
JUMPING MICE

JUMPING MICE – Superficially look like the unrelated kangaroo rats with their long tails and large feet. These nocturnal mice have grooved upper incisors and a tricolored pelage with distinct colors on the back, sides, and belly. Sexes are similar. Not trapped in winter because all species hibernate for most of the cold months.

WOODLAND JUMPING MOUSE *Napaeozapus insignis* 210–255mm, 125–160mm, 15–30g

Redder than the Meadow Jumping Mouse with a white tail tip. Pelage is smooth (not grizzled) and the color becomes more orangish toward southern portion of range. No whitish border on ears. Only three cheek teeth on each side of upper jaw. Feeds on fungi, insect larvae, and fruit. Found only in or along wooded areas; rarely in open habitats.

MEADOW JUMPING MOUSE *Zapus hudsonius* 180–235mm, 100–135mm, 12–30g

Yellowish mouse with a bicolored tail that lacks a white tip. Four cheek teeth. Yellow fur is less grizzled than in the Western Jumping Mouse, with a more distinctly bicolored tail and no white fringe on the ear. Slightly smaller than other *Zapus*. Primarily nocturnal, and escapes by making short leaps; shorter than those of the Woodland Jumping Mouse. Colorado and Wyoming subspecies (*Z. h. preblei*) is threatened. Found in grassy or weedy fields.

WESTERN JUMPING MOUSE *Zapus princeps* 215–245mm, 130–150mm, 18–25g

Pelage grizzled dorsally, with pale yellowish-buff lateral line. Ears have a slight whitish border; tail less distinctly bicolored. Found in high mountain meadows and in wetter streamside habitats.

PACIFIC JUMPING MOUSE *Zapus trinotatus* 210–250mm, 110–155mm, 20–30g

Brightly contrasting colors on back with a brownish (not white) border to the ears. Pelage tricolored, back dark brown, sides yellowish orange, and belly whitish. Distinct area of buff hair forms a lateral line. Fall pelage paler and less contrasting. Tail sparsely haired. Feeds mainly on grass seeds, and leaves characteristic piles of stem cuttings. Found in wet meadow habitats of the Pacific Northwest.

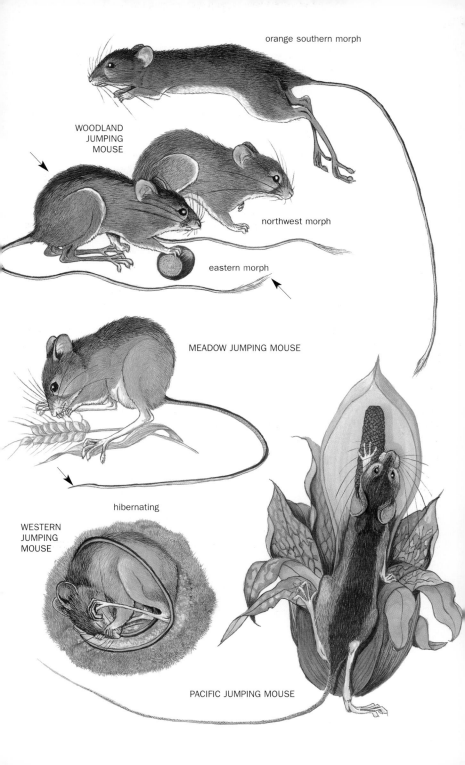

orange southern morph

WOODLAND
JUMPING
MOUSE

northwest morph

eastern morph

MEADOW JUMPING MOUSE

hibernating

WESTERN
JUMPING
MOUSE

PACIFIC JUMPING MOUSE

PLATE 44
WOODRATS 1

WOODRATS – Also known as packrats, these soft-furred, nocturnal rats have long tails and relatively large ears. Characters of the tail or throat often distinguish species.

EASTERN WOODRAT *Neotoma floridana* ♂ 305–450mm, 130–180mm, 220–385g; ♀ 300–400mm, 130–170mm, 175–260g

A medium-sized woodrat, grayish-brown dorsally and white ventrally. Differs from Allegheny Woodrat primarily in having a skull with a maxillo-vomerine notch, and by range. Worn summer pelage may appear cinnamon-orange. Head becomes grizzled in older animals. Larger than Mexican Woodrat, with throat and breast hair that is white to base. Paler in color than Southern Plains Woodrat. Dens constructed of sticks, often around large rock or log. Key Largo subspecies, *N. f. smalli*, is Endangered. Found primarily in wooded areas, but also in hedgerows and rocky outcrops in grasslands.

ALLEGHENY WOODRAT *Neotoma magister* ♂ 370–465mm, 145–215mm, 230–485g; ♀ 350–445mm, 140–210mm, 230–455g

Only woodrat in range. Slightly larger, and with slightly hairier tail than Eastern Woodrat; skull needed for certain identification. A large woodrat, cinnamon brown with gray dorsally and gray or white ventrally. Juveniles are grayer; summer animals are darker. Does not build complex stick dens. Rare and declining, apparently because of a disease transmitted from large Northern Raccoon populations. Found along cliffs, caves, and talus slopes of the Allegheny Mountains.

MEXICAN WOODRAT *Neotoma mexicana* 290–415mm, 105–205mm, 150–255g

Smallish woodrat with sparsely haired tail and throat hairs gray at base. Back may be grayish brown or rufous brown; white below with a bicolored tail. Color is less gray than Southern Plains Woodrat. A dusky line usually borders the mouth. May build stick dens; more typically uses crevices, tree cavities, or buildings for nests. Found on rocky outcrops or cliffs and rocky slopes in mountainous areas and woodlands.

SOUTHERN PLAINS WOODRAT *Neotoma micropus* ♂ 335–410mm, 130–175mm, 210–315g; ♀ 310–380mm, 130–165mm, 180–275g

Steel-gray back with throat hairs white to base. Bicolored tail is relatively short and sparsely haired. Fur soft, dense, darker than in the Eastern Woodrat, and less colorful than the Mexican Woodrat. Belly is pale gray; throat and chest are white. Feet are white. Dens are built from sticks, manure, and cactus joints around desert vegetation. Found on rocky hillsides and grassy lowlands with desert scrub vegetation.

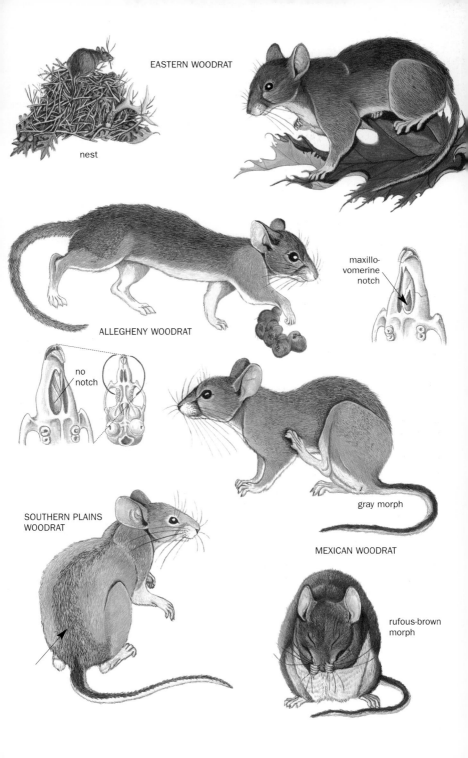

EASTERN WOODRAT

nest

ALLEGHENY WOODRAT

maxillo-
vomerine
notch

no
notch

gray morph

SOUTHERN PLAINS
WOODRAT

MEXICAN WOODRAT

rufous-brown
morph

PLATE 45
FURRY-TAILED WOODRATS

WHITE-THROATED WOODRAT *Neotoma albigula* 280–400mm, 75–185mm, 135–285g

Throat hairs are white to base. This medium-sized woodrat has a brown back and a bicolored tail. Tail has long hairs, but is not bushy. The soft short pelage often has blackish hairs interspersed with the brown. Ears are relatively long and feet are white. Large stick dens are constructed of cactus and woody stems of mesquite, juniper, and other shrubs. Found in wide variety of habitats ranging from low desert to rocky slopes on mountainsides.

STEPHENS'S WOODRAT *Neotoma stephensi* 275–310mm, 115–150mm, 115–180g

Small woodrat with dusky-gray pelage and a semi-bushy tail. Western animals are darker, with throat hair dark to base and a nearly black tail; eastern animals are lighter in color with throat hairs that are white at the base and a tail with fewer black hairs. Belly is creamy in color. Dusky dorsal color extends to just below ankle, feet are white. Tail is bushier than in all woodrats except the Bushy-tailed Woodrat. Dens are almost always found at the base of junipers. Also occurs in ponderosa pine, agave, and cactus habitats.

BUSHY-TAILED WOODRAT *Neotoma cinerea* ♂ 310–470mm, 130–225mm, 180–585g; ♀ 275–410mm, 120–195mm, 165–370g

Large woodrat with bushy, squirrel-like tail. Back ranges from pale gray to blackish brown and belly is buff to whitish; animals in cooler climates have darker coloration. Fur is long and dense. Whiskers are long. Some subspecies have white hair along mid-ventral line, others have white throats. Dens are haphazard accumulations of sticks, bones, and other material, often dropped on cracks and crevices. Urination spots often conspicuous. Found in wooded, mountainous habitats in the Northwest.

WHITE-THROATED WOODRAT

dark-colored western morph
of Stephens's Woodrat

STEPHENS'S WOODRAT

light-colored
eastern morph

dark morph   BUSHY-TAILED
WOODRAT

light morph

PLATE 46
WOODRATS AND *RATTUS*

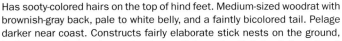

DUSKY-FOOTED WOODRAT *Neotoma fuscipes* 335–470mm, 160–240mm, 205–360g

Has sooty-colored hairs on the top of hind feet. Medium-sized woodrat with brownish-gray back, pale to white belly, and a faintly bicolored tail. Pelage darker near coast. Constructs fairly elaborate stick nests on the ground, in the vegetation, and on rocky slopes. Subspecies *N. f. riparia* from San Joaquin Valley, California, is Endangered. Found mainly in scrub and woodland communities.

ARIZONA WOODRAT *Neotoma devia* ♂ 260–310mm, 110–150mm, 100–130g; ♀ 260–300mm, 115–135mm, 95–125g

Has very large petal-like ears and dark gray throat fur. Small body size with pale coloration. Belly is white. Range restricted to Arizona. Smaller and paler in color than the Desert Woodrat. Builds smallish dens at the base of cliffs and on rocky outcroppings, and covers them with cholla cactus. Found in both low deserts and mountain areas south of Grand Canyon and east of Colorado River.

DESERT WOODRAT *Neotoma lepida* 225–385mm, 95–190mm, 130–160g

Feet are white, ears are large, and throat hairs are dark. Smallish woodrat, pale gray-brown dorsally, with a distinctly bicolored tail. Coastal animals somewhat smaller than those from interior mountains and valleys. Larger and darker than the Arizona Woodrat. Builds sloppy stick nests around cacti, or on rocky crevices. Found in desert scrub and coastal sage scrub habitats.

HOUSE RAT *Rattus rattus* 325–455mm, 160–255mm, 115–350g

Scaly, sparsely haired, uniformly dark tail that is longer than the head and body. This dark rat has a brownish-gray back and a grayish belly (not white like woodrats). Introduced from Europe with earliest colonists. Widespread in south and coastal areas. Frequents urban areas, buildings, and warehouses.

BROWN RAT *Rattus norvegicus* 315–460mm, 120–215mm, 195–485g

Similar to House Rat, but with tail shorter than head and body. Scaly tail and dark belly distinguish it from woodrats. A later introduction that has outcompeted House Rats in many areas. Widespread from southern Canada throughout United States in both urban and old field habitats.

DUSKY-FOOTED WOODRAT

nest

ARIZONA WOODRAT

dark throat patch

DESERT WOODRAT

HOUSE RAT

BROWN RAT

PLATE 47
GRASSHOPPER MICE AND RICE RATS

GRASSHOPPER MICE – *ONYCHOMYS* – These are among the most predaceous rodents, feeding mainly on insects and scorpions at night. They are characterized by a relatively short tail, and have a loud, pure tone vocalization that is audible to humans. The three species are nearly identical, and best distinguished by geographic range.

NORTHERN GRASSHOPPER MOUSE *Onychomys leucogaster*
120–190mm, 30–60mm, 25–50g

Large grasshopper mouse with relatively short tail. Back may be grayish or cinnamon-buff; underside is white. Tail tip may be white. Found in arid and semiarid habitats throughout western North American shrub steppes and grasslands.

SOUTHERN GRASSHOPPER MOUSE *Onychomys torridus*
130–160mm, 40–60mm, 20–40g

Differs from Chihuahuan Grasshopper Mouse only in genetic details, but ranges do not overlap. Compared with the Northern Grasshopper Mouse, is smaller with a relatively long tail. Grayish or pinkish-cinnamon above; white below. Occurs in arid habitats throughout Sonoran and Mojave deserts, plus chaparral in California (yellow area on map).

CHIHUAHUAN GRASSHOPPER MOUSE *Onychomys arenicola* 120–160mm, 35–55mm, 20–35g

(Not illustrated) Differs from Southern Grasshopper Mouse only in genetic details, but ranges do not overlap. Smaller than the Northern Grasshopper Mouse, with a relatively larger tail. Occurs in low desert-scrub habitats in Chihuahuan Desert, and extending slightly northward (see red area on map of Southern Grasshopper Mouse).

RICE RATS – *ORYZOMYS* – This genus of nocturnal, mostly aquatic rats is diverse in the neotropics (more than 36 species), but only two species reach the United States. Semi-aquatic feeders on seeds, vegetation, and occasionally invertebrates. They are mouselike in appearance, with coarse pelage that is neither bristly nor spiny. They have long tails with visible annulations under sparse hair.

COUES'S RICE RAT *Oryzomys couesi* 390–410mm, 130–140mm, 65–70g

Large brown rice rat with long tail. Back is dark brown, sides are pale brown, and belly is even paler. Builds nests of leaves, twigs, and vines in cattails or small trees over the water. Found only in southern Texas in the United States, but extends well down into Mexico in appropriate marshy habitats.

MARSH RICE RAT *Oryzomys palustris* ♂ 195–260mm, 95–120mm, 45–80g; ♀ 190–255mm, 85–115mm, 40–60g

Small grayish rice rat with tail about equal to head and body length. Has a dark mid-dorsal stripe, the belly and feet are whitish, and the tail is bicolored. Active all year. Makes grapefruit-sized nest of grasses. Some consider the silvery-gray subspecies *O. p. argentatus* of Florida Keys to be full species. The subspecies *O. p. natator* in the lower Florida Keys is Endangered. Found in marshy areas of Atlantic coast and southern United States.

NORTHERN GRASSHOPPER MOUSE

cinnamon
morph
singing

SOUTHERN
GRASSHOPPER
MOUSE

gray morph

COUES'S RICE RAT

MARSH RICE RAT

O. p. argentatus

PLATE 48
CENTRAL AND EASTERN TINY MICE

NORTHERN PYGMY MOUSE *Baiomys taylori* 87–123mm, 34–53mm, 6–10g

North America's smallest rodent. Varies from reddish brown through gray to almost black above, and white to creamy buff or gray below. Tail is covered with short hairs, and may be uniformly gray or paler ventrally. Incisors are ungrooved. Uses a variety of habitats including prairie, mixed-desert shrub, post oak savanna, and pine oak forests.

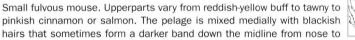

HARVEST MICE – Mice in this genus are distinguished by their small size and the single groove on each upper incisor.

FULVOUS HARVEST MOUSE *Reithrodontomys fulvescens* 134–189mm, 73–116mm, 6–25g

Small fulvous mouse. Upperparts vary from reddish-yellow buff to tawny to pinkish cinnamon or salmon. The pelage is mixed medially with blackish hairs that sometimes form a darker band down the midline from nose to tail. Underparts vary from white to gray, often tinged with buff or pale pinkish cinnamon. Tail is brown to dark brown above and only indistinctly bicolored. Feet are grayish white to buff-white and ears are varying shades of brown, often with a tawny or reddish-yellow tinge on the inner surface. The pelage is coarser than in other harvest mice, with a streaked salt-and-pepper effect from the black guard hairs. Larger and brighter than Eastern Harvest Mouse. Less gray than Plains Harvest Mouse with a relatively shorter tail that is not bicolored. Often trapped near rocky outcrops and cacti. Constructs baseball-size nests of grasses and sedges in vegetation off the ground. Uses grassy fields with shrubs, especially mesquite grassland, grassland, pine-grass ecotones, and grass-brush habitats.

EASTERN HARVEST MOUSE *Reithrodontomys humulis* 107–128mm, 45–60mm, 10–15g

Small, brown mouse with a relatively short tail. Upper parts are rich brown, sometimes faintly washed with gray, and with a dark mid-dorsal stripe. Sides are paler than the dorsum, with an obvious lateral line present. Underparts are ash-colored and often have a cinnamon or pinkish wash. Tail is slender, sparsely furred, bicolored, and shorter than the head and body. Ears are fuscous or fuscous-black in color, and feet are grayish white. Similar to *R. fulvescens*, but tail is shorter and more bicolored. Uses broom sedge, grassy or weedy areas, tangled patches of briar, roadside ditches, brackish meadows, and wet bottomlands.

PLAINS HARVEST MOUSE *Reithrodontomys montanus* 54–146mm, 20–69mm, 6–13g

Small, grayish mouse with a diffuse dark stripe down the middle of its back and a short bicolored tail. Remainder of fur on back is grayish brown and the underside is white. Tail is shorter than head and body, bicolored, with a sharp dark line above and paler color below. Tail is sparsely haired, but does not appear scaly. There is a bright tuft in front of the ears and a spot behind them. Similar to Eastern Harvest Mouse, but larger and paler, with paler ears. Uses grassy areas in the Great Plains.

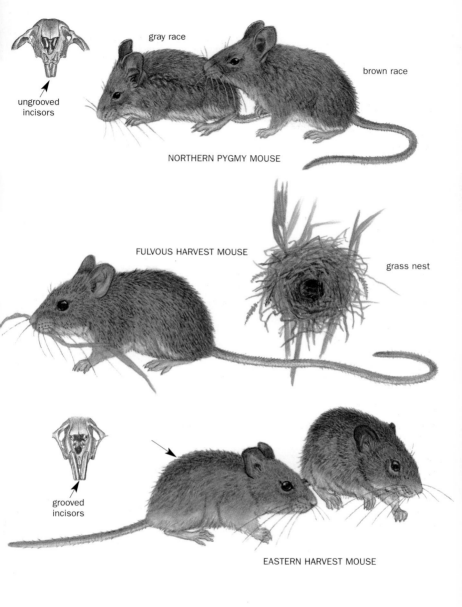

gray race

brown race

ungrooved
incisors

NORTHERN PYGMY MOUSE

FULVOUS HARVEST MOUSE

grass nest

grooved
incisors

EASTERN HARVEST MOUSE

PLAINS HARVEST MOUSE

PLATE 49
HOUSE AND WESTERN TINY MICE

WESTERN HARVEST MOUSE *Reithrodontomys megalotis*
118–170mm, 50–96mm, 8–15g

Small mouse with a relatively long, bicolored tail. Fur is bristly and relatively short. Feet are whitish. The ears are comparatively small and buff to reddish brown. The tail is about as long as the head and body and distinctly bicolored, with relatively long hairs that tend to obscure the scales. Juveniles are grayish brown; subadults are brighter than juveniles but duller than adults. The thick winter pelage is paler and the tail is more distinctly bicolored. Some animals east of the Mississippi River and in the San Francisco Bay area have a buff pectoral spot. Larger than Plains Harvest Mouse, with longer hair and a more distinctly bicolored tail. Larger than Eastern Harvest Mouse, with a relatively longer tail. Has a less pointed, more bicolored tail than Saltmarsh Harvest Mouse. Uses open mesa habitats dominated by herbaceous vegetation and dense litter, such as prairies, meadows, overgrown pastures, stream valleys, and estuarine marshes.

SALTMARSH HARVEST MOUSE *Reithrodontomys raviventris*
118–175mm, 56–95mm, 7–15g

A small reddish mouse from the San Francisco Bay area. The subspecies from the south end of the bay (*R. r. raviventris*) has dark, cinnamon-colored back with a tawny lateral line and a pale to cinnamon belly. Mice from the north end of the bay (*R. r. halicoetes*) have paler upperparts with no lateral line and a white belly. Similar to Western Harvest Mouse, but with a more unicolored tail. Known for its docile behavior, this species is endangered and restricted to salt marshes bordering San Francisco Bay.

HOUSE MOUSE *Mus musculus* 130–200mm, 65–100mm, 18–23g

Small, nearly unicolored grayish-brown mouse with dark underparts. Tail is dusky and unicolored. Most likely to be confused with the Harvest Mouse, which has grooved incisors and typically has reddish back and lighter belly, or *Peromyscus*, which has white underparts. This cosmopolitan, introduced species is most common around human habitations, but may inhabit old fields and nearby disturbed habitats as well.

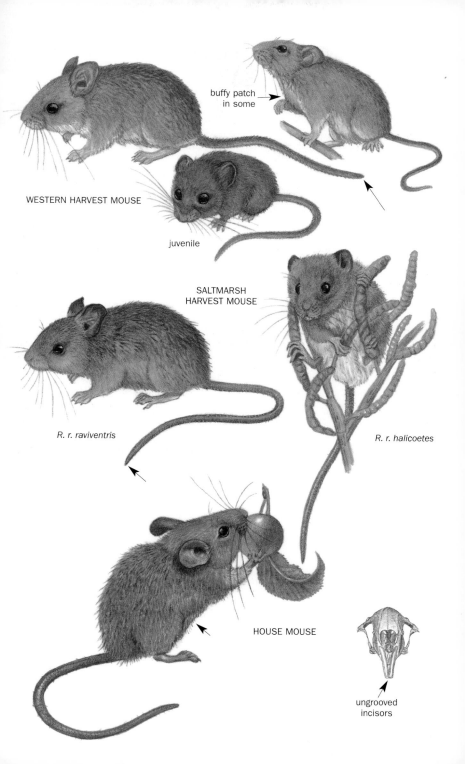

buffy patch
in some

WESTERN HARVEST MOUSE

juvenile

SALTMARSH
HARVEST MOUSE

R. r. raviventris

R. r. halicoetes

HOUSE MOUSE

ungrooved
incisors

PLATE 50
NORTHERN *PEROMYSCUS*

PEROMYSCUS MICE – Mice in this diverse group are characterized by brownish backs, white bellies, and long tails. Their large eyes and ears are adaptations to their nocturnal habits. The different species are distinguished by details of their measurements, and body and tail coloration.

WHITE-FOOTED DEERMOUSE *Peromyscus leucopus* 150–205mm, 65–95mm, 15–25g

Small, grayish to brownish mouse, frequently with dark stripe along mid-back, and white underparts. Tail sparsely haired and indistinctly bicolored, tuft of hairs on tip less than 5mm long. Very similar to other species of *Peromyscus*: North American Deermouse typically has a distinctly bicolored tail (although not in the northeastern United States); Cotton Deermouse has a longer hind foot; Oldfield Deermouse is smaller and pale cinnamon to almost white in color, and Brush, White-ankled, Piñon, Saxicolous, Northern Rock, and Texas Deermouse all have longer tails. Favors warm dry forests at low to mid-elevations.

NORTH AMERICAN DEERMOUSE *Peromyscus maniculatus* 120–225mm, 50–125mm, 10–30g

The most widespread, geographically and ecologically variable mouse in North America. Has large black bulging eyes, relatively large, naked ears, fine, smooth-lying fur, and white feet. The well-haired, sharply bicolored tail is tipped with a tuft of short, stiff, hairs something like a watercolor brush. Juveniles are gray, subadults are yellowish brown. The two main forms include a long-tailed, large-eared forest dweller and a short-tailed, small-eared, open-country form. Generally has darker, richer colors in humid regions and paler, drabber colors in arid regions. Differs from White-footed Deermouse by having a distinctly bicolored tail that is about as long as the head and body length; from Northwestern Deermouse by having a shorter tail and foot; from Brush and White-ankled Deermouse by having a shorter tail; from California Deermouse by being smaller; from Cactus and Merriam's Deermouse by being darker in color with more hair on tail and more distinctly bicolored tail; from Canyon Deermouse by having darker color, shorter tail, and shorter, less soft pelage; and from Piñon and Northern Rock Deermouse by having smaller ears. Found in almost all habitat types within its range.

NORTHWESTERN DEERMOUSE *Peromyscus keeni* 181–236mm, 92–114mm, 10–30g

Large mouse with dark colors, long tail, and dense fur. Similar to North American Deermouse, but significantly larger, and with a longer tail. Occupies habitats ranging from coastal lowlands to sub-alpine forests. Tends to be more common in edge habitats, and less so in dense coniferous forests.

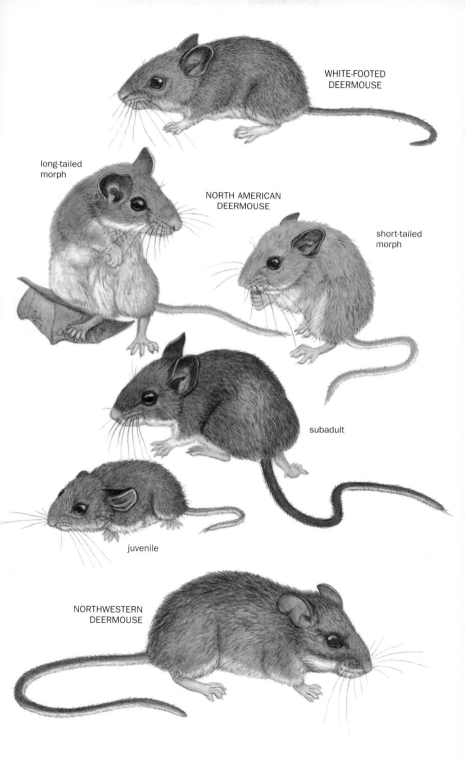

WHITE-FOOTED
DEERMOUSE

long-tailed
morph

NORTH AMERICAN
DEERMOUSE

short-tailed
morph

subadult

juvenile

NORTHWESTERN
DEERMOUSE

PLATE 51
SOUTHEASTERN *PEROMYSCUS*

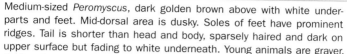

COTTON DEERMOUSE *Peromyscus gossypinus* 142–206mm, 55–97mm, 17–46g

Medium-sized *Peromyscus*, dark golden brown above with white underparts and feet. Mid-dorsal area is dusky. Soles of feet have prominent ridges. Tail is shorter than head and body, sparsely haired and dark on upper surface but fading to white underneath. Young animals are grayer. Larger than Oldfield and North American Deermouse. Also larger than White-footed Deermice of the same age. The subspecies from Key Largo (*P. g. allapaticola*) is endangered. Prefers somewhat wet habitats, especially bottomland hardwood forests, hammocks, and swamps.

OLDFIELD DEERMOUSE *Peromyscus polionotus* 110–150mm, 40–60mm, 10–15g

Small *Peromyscus* whose inland forms are fawn-colored or brownish gray dorsally, slightly darker along the midline with hairs that are slate-gray at the base. Underparts are white with hairs pigmented at the base, and the bicolored tail has a dark dorsal stripe. Beach forms are paler and have underparts that are white to the base of the hairs, and a dark tail stripe that is reduced or absent. Many coastal subspecies are Endangered because of the declining health of the Gulf states' coastal dune ecosystem. Inhabits open, sandy habitats in early successional stages of abandoned fields, and grassy dunes along the coast.

FLORIDA DEERMOUSE *Podomys floridanus* 178–220mm, 80–101mm, 27–47g

Large mouse; brownish above, and white below, with orange lateral areas separating the two. Larger than other *Peromyscus* in range, with bigger ears and feet. Has orange color on cheeks, shoulders, and sides. Has a distinctive skunklike odor. Known to frequent burrows of gopher tortoises. Uses sandy uplands with prickly pear cactus and longleaf pines.

GOLDEN MOUSE *Ochrotomys nuttalli* 140–190mm, 67–97mm, 18–27g

Medium-sized mouse with golden upperparts and creamy-colored underparts and feet. Pelage is very soft and hairs are very fine. Attractive and relatively docile, but rarely encountered in or near human habitations. Degree of brightness of color varies by subspecies. Uses densely forested lowlands and floodplain communities, often including honeysuckle, greenbrier, or other vines.

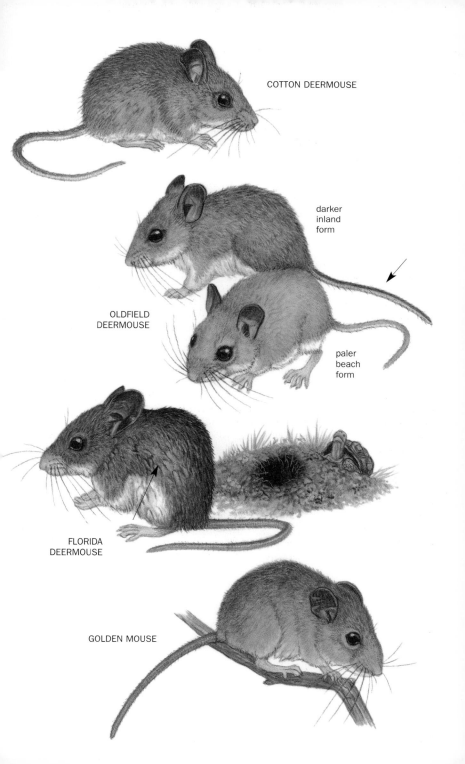

COTTON DEERMOUSE

darker
inland
form

OLDFIELD
DEERMOUSE

paler
beach
form

FLORIDA
DEERMOUSE

GOLDEN MOUSE

PLATE 52
FAR WEST PEROMYSCUS

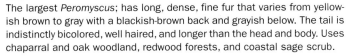

CALIFORNIA DEERMOUSE *Peromyscus californicus* 220–285mm, 117–156mm, 33–55g

The largest *Peromyscus*; has long, dense, fine fur that varies from yellowish brown to gray with a blackish-brown back and grayish below. The tail is indistinctly bicolored, well haired, and longer than the head and body. Uses chaparral and oak woodland, redwood forests, and coastal sage scrub.

CANYON DEERMOUSE *Peromyscus crinitus* 162–191mm, 79–118mm, 13–23g

Small to medium-sized mouse with relatively long ears. The tail is as long as the head and body and thinly haired with a distinct "pencil" of hairs at the tip. Dorsal pelage is long and soft, varying in color from brown to pale buff. Venter is white, sometimes with a buff pectoral or anal patch. Feet are white. Tail is more tufted and densely haired than in Merriam's or Cactus Deermouse. Uses arid grasslands and shrublands, and slickrock deserts of west, where it often is found on bare rock surfaces.

CACTUS DEERMOUSE *Peromyscus eremicus* 169–218mm, 92–117mm, 18–40g

Medium-sized mouse with long, soft, silky pelage and large ears. Body is pale gray washed with reddish brown above and whitish on the underparts. The head is grayish, and the lateral line is pale yellowish buff. The tail is markedly longer than the head and body, bicolored, finely ringed, thinly haired, and lacking a prominent tuft on the tip. Ears are large, thin, and barely covered with fine hair. Soles of the hind feet are naked to the heel. Animals on old lava flows are darker. Smaller and grayer than the California Deermouse with a lateral line and less hair on the tail. Usually smaller than the Merriam's Deermouse and lighter in color, with white (not creamy) underparts and without a cinnamon-colored pectoral patch. Lacks the densely-haired tail with tufted tip of the Brush or Canyon Deermouse. Has smaller ears than the Piñon Deermouse. Uses low desert areas and rocky foothills with scattered vegetation and sandy soils. Prefers rocky areas.

MERRIAM'S DEERMOUSE *Peromyscus merriami* 185–225mm, 95–120mm, 20–30g

Small to medium-sized mouse, somewhat darker above than the Cactus Deermouse, with creamy underparts. The tail is nearly naked and longer than the head and body. The ears are relatively small and without white rims. Usually larger than the Cactus Deermouse, darker on the back, with a creamy (not white) venter. Has a less densely haired tail than the Canyon Deermouse without a tufted tip. Occurs in bosques – dense thickets of mesquite, often mixed with cholla, prickly pear, palo verde, vines, and grasses.

CALIFORNIA
DEERMOUSE

CANYON DEERMOUSE

CACTUS DEERMOUSE

MERRIAM'S DEERMOUSE

PLATE 53
PEROMYSCUS TRUEI AND *BOYLII* GROUPS

PIÑON DEERMOUSE *Peromyscus truei* 171–231mm, 76–123mm, 15–50g

Medium-sized mouse with large ears and a hairy tail. Ears are longer than the hind feet, and relatively larger than other *Peromyscus*. The fur is long and silky, ranging from pale yellowish brown to dark grayish brown above, and the underparts and feet are white. The tail has a dark dorsal stripe and is tipped with long hairs. Favors rocky slopes with pinyon pine and juniper.

NORTHERN ROCK DEERMOUSE *Peromyscus nasutus* 194–198mm, 98–102mm, 24–32g

Fairly large, gray-brown mouse with a long, hairy, bicolored tail and large ears. The underparts are paler. Quite similar to *P. boylii*, but with slightly larger ears (17–28mm). Smaller than *P. gratus* with slightly longer feet (22–28mm) and ears. The tail sheath is much more likely to break in *P. nasutus* than in *P. boylii*. Tail has more fur than most other *Peromyscus*. Uses boulder-strewn regions in pinyon-juniper and oak woodland zones.

SAXICOLOUS DEERMOUSE *Peromyscus gratus* 171–231mm, 76–124mm, 19–33g

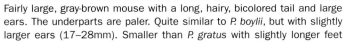

Has a brownish to brownish-black back and a long tuft of hairs on the tail tip. The underparts and hind feet are whitish. The tail is longer than the head and body, brownish on top, whitish underneath, and covered with short hairs except for long hairs on tip. Fur is darker in populations living on lava flows. Slightly larger than Northern Rock Deermouse, with a slightly longer tail, and slightly shorter hind feet (20–26mm) and ears (17–23mm). Uses a variety of rocky areas.

TEXAS DEERMOUSE *Peromyscus attwateri* 187–218mm, 96–112mm, 25–35g

Medium-sized *Peromyscus* with large hind feet and ankles that are usually dark or dusky, and ears that are medium-sized compared to other *Peromyscus*. The tail is bicolored, well tufted, and equal in length to head and body. Uses cedar glade, juniper-grass, and oak-juniper forests.

BRUSH DEERMOUSE *Peromyscus boylii* 175–210mm, 89–115mm, 22–36g

Has unique broad, bright orange lateral line extending from the cheek to the hindquarters. The back is medium brown and the sides paler brown, grading to white or cream on underparts. Ankles are dusky gray. Ears and hind feet are similar in size. Tail is bicolored, longer than the head and body, well haired, and tufted at end. Has smaller ears (16–20mm) than the Northern Rock Deermouse. Uses rock outcroppings and brushy or forested areas in elevations more than 2000m. Rock ledges, boulders, brush piles, and fallen trees are typical of its habitat.

WHITE-ANKLED DEERMOUSE *Peromyscus pectoralis* 185–219mm, 92–117mm, 24–39g

Medium in size, with ears shorter than the hind feet, ankles that are usually white, and a tail that is hairy and coarsely ringed. The tail is equal to or decidedly longer than the head and body. Similar to Texas Deermouse, but slightly paler, with a more distinctly bicolored tail and smaller hind feet (less than 24mm). Uses rocky situations in arid mountain regions and in Texas Hill Country.

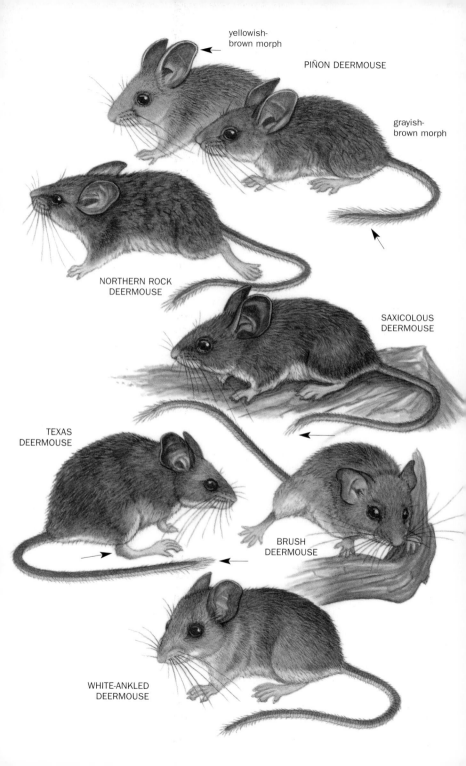

yellowish-
brown morph

PIÑON DEERMOUSE

grayish-
brown morph

NORTHERN ROCK
DEERMOUSE

SAXICOLOUS
DEERMOUSE

TEXAS
DEERMOUSE

BRUSH
DEERMOUSE

WHITE-ANKLED
DEERMOUSE

PLATE 54
SIGMODON

COTTON RATS – *SIGMODON* SPP. – Vaguely volelike in general appearance but larger, cotton rats also make runways in grassy areas. These medium-sized rats have stocky bodies with somewhat harsh fur. The three central digits on the hind foot are larger than the other two. The generic name refers to the S-shaped pattern of the cusps of the cheek teeth.

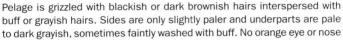

HISPID COTTON RAT *Sigmodon hispidus* 224–365mm, 81–166mm, 110–225g

Pelage is grizzled with blackish or dark brownish hairs interspersed with buff or grayish hairs. Sides are only slightly paler and underparts are pale to dark grayish, sometimes faintly washed with buff. No orange eye or nose ring. Smaller than the Arizona Cotton Rat, with smaller feet (32–34mm). Tail short, scaly, and sparsely haired. Uses habitats with grasses especially little bluestem and bushy beardgrass.

YELLOW-NOSED COTTON RAT *Sigmodon ochrognathus* 132–264mm, 80–114mm, 51–106g

Small cotton rat with a yellowish patch on each side of the nose. Back is muddy gray and underparts are silvery or whitish. Color of the inside of the pinna of the ear the same as the dorsum. Dorsal guard hairs and underhairs are grayish with a tinge of yellow, whereas those of the venter are grayish white. Small hind feet (less than 30mm) are buff-gray, and tail is hairy and blackish above, grayish below. Uses grassy patches in montane situations, especially dry rocky slopes.

ARIZONA COTTON RAT *Sigmodon arizonae* 200–349mm, 85–156mm, 125–211g

Look like grizzled rats with scaly tails. They are similar to, but slightly larger than Hispid Cotton Rats with a hind foot larger than 34mm. Common in grassy areas around ponds, drainages, riparian areas, irrigated fields, and weedy or brushy areas.

TAWNY-BELLIED COTTON RAT *Sigmodon fulviventer* 223–270mm, 94–109mm, 200–222g

The largest *Sigmodon* has a back that is pepper and salt in color and underparts that are washed with buff. Tail is dark and usually covered with just enough hairs to hide the scales, which are smaller than those of the Hispid Cotton Rat. Uses grass and grass-shrub habitats.

HISPID COTTON RAT

YELLOW-NOSED COTTON RAT

ARIZONA COTTON RAT

TAWNY-BELLIED COTTON RAT

HISPID COTTON RAT

PLATE 55
ARBORIMUS VOLES

ARBORIMUS VOLES – These small, reddish voles almost never descend to the ground from their arboreal homes. Therefore, they are rarely seen and poorly known. Long thought to be a subgenus of heather voles, these tree voles differ in having longer tails, smaller ears, longer feet, and wider and shorter teeth.

RED TREE VOLE *Arborimus longicaudus* 158–206mm, 60–94mm, 25–47g

Slightly larger than White-footed Vole, with pelage that is thick, long, and soft. Back is red and lacks a medial stripe. Animals from the northern Oregon coast are larger, with brownish-red backs and pale gray undersides; southern and inland animals are brighter colored with undersides washed in reddish orange. Tail is long, hairy, and black to brown. Eyes are small and ears are pale and hairless. Distinguished from Sonoma Tree Vole by not being as brightly colored and by having nasal bones that extend beyond the maxillaries. Spends most of time in the tops of tall conifers, eating needles of Douglas fir.

SONOMA TREE VOLE *Arborimus pomo* 158–187mm, 60–83mm, 25–47g

Like the other *Arborimus*, but with reddish fur that is gray at the base. Many hairs are tipped slightly with black. Belly is white (with gray base to hairs) and often washed with reddish orange. Skin on tail is black and covered with reddish fur. Tail is well haired, thick, and more than half the head and body length. Ears are almost hairless. More brightly colored than Red Tree Vole, with nasal bones that do not extend beyond the maxillaries. Very arboreal in coniferous trees.

WHITE-FOOTED VOLE *Arborimus albipes* 149–182mm, 57–75mm, 17–29g

Small wooly mouse with white feet and fur that is long, soft, and a warm brown color. Belly is gray and sometimes washed with pale brown. Long, thinly-haired tail is black on top and white below. Feet are usually white on top, and ears, which are usually hidden by the fur, are hairless. May be the most terrestrial of the *Arborimus*, found in riparian alder habitats along small streams.

Red Tree Vole with nest

northern Oregon

RED TREE VOLE

inland southern Oregon

Red Tree Vole

Sonoma Tree Vole

SONOMA TREE VOLE

WHITE-FOOTED VOLE

PLATE 56
RED-BACKED VOLES

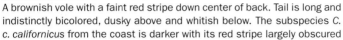

RED-BACKED VOLES – *CLETHRIONOMYS* SPP. – Aptly named, with a chestnut-brown to reddish-brown stripe on the back that grades to dark gray or buff-gray on the sides and belly. Pelage is long and soft in winter and short and coarse in summer.

WESTERN RED-BACKED VOLE *Clethrionomys californicus*
121–165mm, 34–56mm, 15–40g

A brownish vole with a faint red stripe down center of back. Tail is long and indistinctly bicolored, dusky above and whitish below. The subspecies *C. c. californicus* from the coast is darker with its red stripe largely obscured by black hair. Inland animals are typically paler. Dorsal stripe not as obvious as in the Southern Red-backed Vole. Tail is longer and more bicolored than in the Northern Red-backed Vole. Favors western coniferous forests with little undergrowth. Forages mostly under surface of the forest floor.

SOUTHERN RED-BACKED VOLE *Clethrionomys gapperi*
116–172mm, 30–50mm, 6–42g

Brilliantly colored, can easily be distinguished by broad, reddish band running from forehead to rump. Nose, sides of head, and body are gray, often with a yellowish cast. Belly colors range from silvery white to pale yellowish, and the tail is bicolored. There are two distinct color phases in northern and eastern subspecies with either bright reddish or more grayish-brown dorsal bands. This dorsal stripe is brighter and more distinct than in other *Clethrionomys* species. Feet are pale drab gray. Uses natural runways along and beneath logs, rocks, and roots of trees. Found in mesic coniferous, deciduous, and mixed forests with abundant litter of stumps, rotting logs, and exposed roots.

NORTHERN RED-BACKED VOLE *Clethrionomys rutilus*
127–161mm, 30–48mm, 23–40g

This species is very similar to the Southern Red-backed Vole, but has a duller red back with slightly more red hairs on the sides. The tail is relatively thick and short. Occurs in both taiga and tundra habitats, all essentially north of the range of the Southern Red-backed Vole.

darker morph

paler inland morph

WESTERN RED-BACKED VOLE

grayish-brown morph

reddish morph

SOUTHERN RED-BACKED VOLE

Western Red-backed Vole

NORTHERN RED-BACKED VOLE

PLATE 57
COASTAL *MICROTUS*

MICROTUS VOLES – The most common and diverse of the voles, typically live in grasslands where they form runways between their nests and feeding areas. Voles are stout-bodied mice with long, loose pelage and short tails. The ears are so short and rounded as to be almost concealed by the pelage.

CALIFORNIA VOLE *Microtus californicus* 139–207mm, 38–68mm, 30–81g

Medium-sized vole with tail of moderate length and belly fur that is pale with gray at the base. Grizzled brownish with scattered black hairs above. Long tail is bicolored. Feet are pale. Occurs at lower elevations than Montane Vole; has shorter and less bicolored tail than Long-tailed Vole; larger than Creeping Vole, and has six plantar tubercles; slightly larger than Townsend's Vole. The subspecies near Inyo County, California (*M. c. scirpensis*), is Endangered. Prefers low-elevation grasslands and wet meadows, but also found in coastal wetlands and open oak savannas with good ground cover.

GRAY-TAILED VOLE *Microtus canicaudus* 140–168mm, 32–45mm, 35–55g

Small but robust vole with large eyes. Yellowish-gray or brown above, venter is grayish white. Feet are gray, and short tail is gray with a brownish dorsal stripe. Has shorter tail and paler fur than Townsend's Vole; has larger eyes and is more robust than Creeping Vole. Uses low-elevation grasslands, including agriculture planted in small grains or legumes.

CREEPING VOLE *Microtus oregoni* 130–153mm, 30–41mm, 17–20g

Small vole with short, dense dorsal fur that is sooty gray to dark brown or almost black with a mixture of yellowish hairs. Belly is a dusky gray to white. Eyes are especially small, tail is short, almost black on top and gray underneath. Small size and tiny eyes distinguish it from other voles. Occupies grassy and herbaceous sites within moist coniferous forests, including recent clearcuts.

TOWNSEND'S VOLE *Microtus townsendii* 169–225mm, 48–70mm, 47–83g

One of the largest voles, with dark brownish fur and large ears. Tail is long, and blackish or brownish. Feet are brownish or blackish with brown claws. Ears are large and broad and extend above the fur. Large size distinguishes it from other voles. Found in wet meadows and marshes with dense grass and sedge cover.

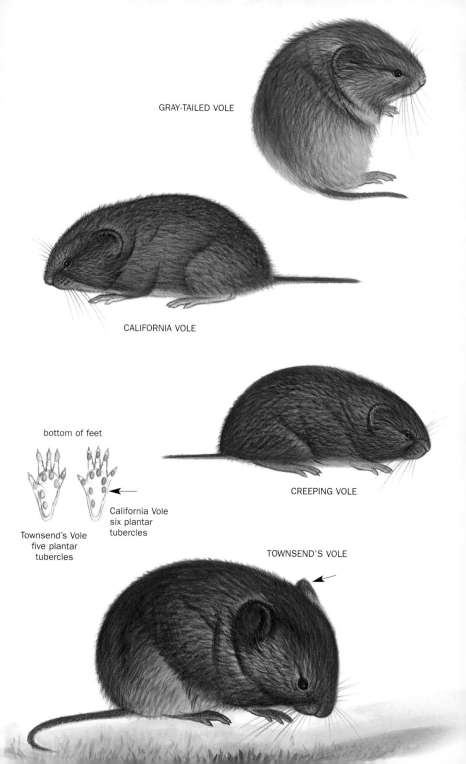

GRAY-TAILED VOLE

CALIFORNIA VOLE

CREEPING VOLE

bottom of feet

Townsend's Vole
five plantar
tubercles

California Vole
six plantar
tubercles

TOWNSEND'S VOLE

PLATE 58
OTHER WESTERN VOLES

MONTANE VOLE *Microtus montanus* 140–220mm, 24–64mm, 18–90g

Small vole, grizzled brown to blackish above, often with buff tint. Belly fur is white to gray, not buffy like Prairie Vole. Moderately long, bicolored tail, but shorter than that of Long-tailed Vole. Feet are dusky or silver gray, darker than in the White-footed Vole, but lighter than in the Meadow Vole. Back is paler than that of Meadow Vole. Adult males often have oily skin glands on hips in breeding season, unlike most other voles. Found in dense woods, wet meadows, and stream sides and, especially, mesic grasslands.

LONG-TAILED VOLE *Microtus longicaudus* 155–202mm, 49–81mm, 36–59g

Small, thick-bodied, grayish vole with long bicolored tail. Feet have six plantar tubercles. Ears are large and haired, and eyes are large. Color of back from ashy gray to brownish gray. Numerous black-tipped hairs occur on the back, but the sides are more grayish. Color and long tail distinguish it from other voles. Occurs in mountain-top habitats with coniferous and hardwood forests, brushy thickets, forest meadow ecotones, and riparian areas.

WATER VOLE *Microtus richardsoni* 234–274mm, 66–98mm, 72–150g

North America's largest vole. Has long pelage that is grayish brown or dark reddish brown on the back, often darkened with black-tipped hairs. Underparts are grayish with a white or silvery-white wash. Bicolored tail is dusky above and grayish below. Has five plantar tubercles on the feet. Large hind feet (more than 23mm) distinguish it from other voles. Semiaquatic, and uses subalpine and alpine meadows close to water, especially swift, clear spring-fed or glacial streams with gravel bottoms, and along the edges of high-elevation ponds.

SAGEBRUSH VOLE *Lemmiscus curtatus* 103–142mm, 16–30mm, 17–38g

Tiny vole with drab buff to ash-gray pelage that is lax, long, and dense. Base of hairs are dark gray. Tail is short, indistinctly bicolored with a dusky line above, and silvery white to buff below. Feet are white or light gray to pale buff, with six plantar tubercles. Lives in colonies in semi-brushy canyons dominated by sagebrush or rabbit brush mixed with bunchgrass.

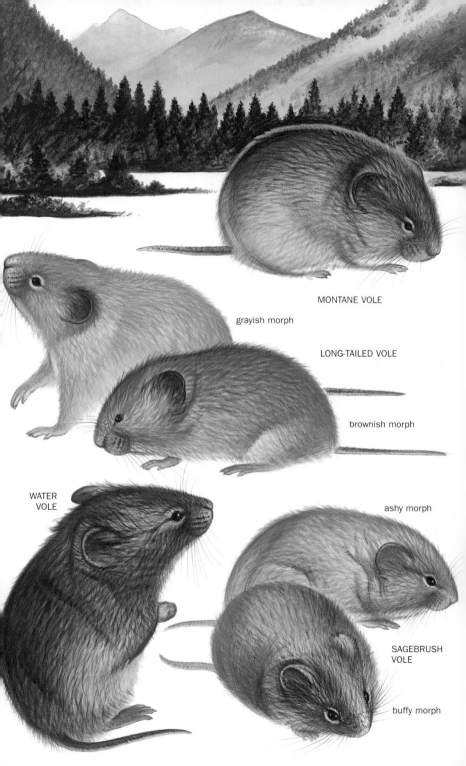

MONTANE VOLE

LONG-TAILED VOLE

grayish morph

brownish morph

WATER
VOLE

ashy morph

SAGEBRUSH
VOLE

buffy morph

PLATE 59
NORTHERN VOLES

TAIGA VOLE *Microtus xanthognathus* 152–226mm, 38–53mm, 85–158g

Large vole with grayish-brown fur and a characteristic yellowish-orange nose, which distinguishes it from the Tundra Vole. Has small, dark, bead-like eyes, small ears, and short tail. Occurs in riparian, boreal, and sphagnum forest habitats near streams and other moist areas.

INSULAR VOLE *Microtus abbreviatus* 136–176mm, 25–32mm, 45–79g

Large, short-tailed vole from Arctic islands. Nearly identical to Singing Vole, but slightly larger. Adults are brownish dorsally, with pale yellowish sides, rump, tips of ears and face, and a buff-colored belly. Restricted to a few offshore Alaskan islands, where it frequents moist, well-drained lowlands and ryegrass areas of beach ridges.

SINGING VOLE *Microtus miurus* 125–168mm, 20–36mm, 22–60g

Smallish, short-tailed vole. Usually quite buff in color on the flanks and venter, with rather enlarged claws. Differs from other voles in combination of short tail, buff venter, and long claws. In late summer, sits in exposed places and makes a metallic churring sound, hence the name Singing Vole. Often associated with willow in well-drained tundra, and extends up into subalpine and alpine zones.

EASTERN HEATHER VOLE *Phenacomys ungava* 122–155mm, 26–41mm, 25–40g

Small vole with grizzled brown fur with a yellowish wash. The tail is short, and the ears are hardly visible above the fur. Ear tips, nose, and rump are usually more tawny or yellowish than those of other voles. Uses open dry country and especially deciduous shrubby habitat such as willow thickets, poplars, or birch meadows.

WESTERN HEATHER VOLE *Phenacomys intermedius* 130–153mm, 26–41mm, 15–41g

Small, short-tailed vole with long and silky fur that is speckled gray to brownish on the back and paler whitish to silver-gray on underparts. The tail is thin, sparsely haired, and distinctly bicolored (dark gray above, white below). Unique with stiff orange hairs in ears. Whiskers reach the shoulders. Differs from other voles in dental characters. Uses boreal habitats including open coniferous forests, riparian areas, and moist alpine and subalpine meadows.

TUNDRA VOLE *Microtus oeconomus* 152–225mm, 30–54mm, 25–80g

Medium-sized for genus with short ears and a short tail. Upper parts range from dusky gray through rich buff to tawny, cinnamon brown, or rusty brown. All color morphs have a mixture of black-tipped hairs. Sides are paler, and underside is white, sometimes washed with dark buff. Tail is bicolored. Larger than Singing Vole, and has broader skull. Slightly larger than Meadow Vole, and lacks rounded posterior loop on second upper molar. Uses moist mountain meadows of the Arctic tundra, especially near streams, lakes, and marshes.

TAIGA VOLE

INSULAR VOLE

SINGING VOLE

WESTERN HEATHER VOLE

EASTERN HEATHER VOLE

ear close up

fulvous morph

TUNDRA VOLE

yellowish morph

PLATE 60
EASTERN VOLES 1

MOGOLLON VOLE *Microtus mogollonensis* 123–144mm, 25–34mm, 18–42g

Small vole with grizzled cinnamon-brown upperparts and buff to cinnamon underparts. Relatively short tail distinguishes it from Meadow Vole and Long-tailed Vole. Females have only four mammae, rather than six or eight as in Prairie Vole and some other voles. Belly is tannish rather than whitish as in Montane Vole. Normally found in grassy meadows in coniferous forests.

PRAIRIE VOLE *Microtus ochrogaster* 130–172mm, 24–41mm, 37–48g

Small vole with long, coarse fur that is grizzled grayish brown from hairs with black and brownish-yellow tips. Sides are slightly paler. Belly is a neutral gray or is washed with whitish or pale cinnamon. Tail is strongly bicolored, and longer than that of Pine Vole and Southern Bog Lemming. Differs from Meadow Vole in having shorter tail, five toe pads, coarser fur, and venter without silver-tipped hairs. Occurs in all types of prairie habitats, plus agricultural areas such as fencerows and fallow fields.

BEACH VOLE *Microtus breweri* 165–215mm, 35–60mm, 45–63g

An insular offshoot of the Meadow Vole, the Beach Vole is slightly larger, paler, and more grizzled. Frequently has a white blaze on forehead or elsewhere on face. Limited to Muskeget Island, Massachusetts, where it prefers beach grass stands, which provide its major food source in addition to shelter.

MEADOW VOLE *Microtus pennsylvanicus* 140–195mm, 33–64mm, 33–65g

Robust vole with a relatively short tail and compressed muzzle. Dull brown above with a gray belly. Immatures slightly darker than adults. Winter pelage is thicker and finer than the sparser, coarser summer coat. Distinguished from other voles by five closed triangles on first lower molar, three transverse loops and no triangles on third lower molar, four closed triangles with a posterior loop on second upper molar, and three closed triangles on third upper molar. The subspecies *M. p. dukecampbelli* known only from Cedar Key, Florida, is Endangered. This is the most prolific mammal on Earth, and occupies moist grassy fields and meadows over much of northern North America.

Prairie Vole

Mogollon Vole

PRAIRIE VOLE

MOGOLLON VOLE

BEACH VOLE

MEADOW VOLE

juvenile

PLATE 61
EASTERN VOLES 2 AND BOG LEMMINGS

WOODLAND VOLE *Microtus pinetorum* 111–139mm, 12–29mm, 14–37g

Tiny reddish vole with reduced eyes and ears and large foreclaws for its semifossorial life. Tail is short and facial vibrissae are well developed. Fine fur is chestnut-colored above and on the sides and paler gray or silvery below. Tail is bicolored, but colors grade gently and no sharp distinct line is noticeable, distinguishing it from other reddish voles. Whiskers are well developed. Incisors are not grooved. Common in eastern forests with good litter cover or grassy areas.

ROCK VOLE *Microtus chrotorrhinus* 140–185mm, 42–64mm, 30–48g

Medium-sized vole with yellowish-orange or pale yellowish wash on snout, and occasionally on rump as well. Smaller than the otherwise similar Taiga Vole, which occurs only to the northwest of the Rock Vole. Uses hardwoods and mixed deciduous coniferous forests with rocks, frequently near streams or other water sources.

BOG LEMMINGS – *SYNAPTOMYS* – Smaller than voles, but with a relatively large head and facial hairs surrounding the snout, which can be erected to make the face appear larger than it really is. Although the head is relatively large, the rostrum is short. Both have grooved incisors, which are sharper in *S. borealis*.

SOUTHERN BOG LEMMING *Synaptomys cooperi* 94–154mm, 13–24mm, 21–50g

Similar to *S. borealis*, but with dark brown above and pale gray on underside. Also has six mammae instead of eight, as in *S. borealis*. Occurs in a wide variety of habitats including clearings in woodlands, grasslands, mixed deciduous/coniferous woodlands, spruce-fir forests, and freshwater wetlands.

NORTHERN BOG LEMMING *Synaptomys borealis* 110–140mm, 17–27mm, 27–35g

Small lemming with grizzled gray to brown pelage on the back and pale gray underneath. Fur has a coarse, ruffled appearance and the short tail is bicolored. Distinguished from *S. cooperi* by having buff-colored hairs at base of the ears, sharper incisors, and more mammae. Occurs in a variety of habitats that are wet with many sedges and grasses including spruce-fir forests, wet meadows, sphagnum bogs, and alpine tundra.

WOODLAND VOLE

ROCK VOLE

SOUTHERN BOG LEMMING

NORTHERN BOG LEMMING

PLATE 62
NORTHERN LEMMINGS

NORTHERN LEMMINGS – These two genera of lemmings inhabit Arctic tundra. In good years, their populations can explode and large numbers of animals can be seen running across the low tundra, dispersing out of overpopulated areas. This has led to the myth of suicidal, cliff-jumping lemmings.

BROWN LEMMING *Lemmus trimucronatus* 130–180mm, 18–26mm, 45–130g

Relatively large lemming with tawny brown to cinnamon on backs and sides, and paler, more buff colors underneath. Older adults may develop a rusty-colored patch on the rump. The body is stout with a blunt muzzle, short tail, and small ears and eyes. Uses subalpine tundra above timberline.

UNGAVA COLLARED LEMMING *Dicrostonyx hudsonius* 125–166mm, 12–16mm, 35–85g

Seasonally variable; in winter they molt to white and develop bifurcated digging claws on forefeet. In summer they are brownish-gray, usually with a dull reddish patch of hair around the ears, in contrast to the more steel-gray pelage of Northern Collared Lemmings. A pale yellowish-orange line on the sides of the body between legs and edge of rump separates the back from the gray belly, and there is usually an indistinct reddish band or collar across the chest. There is a thin black stripe that starts at the tip of the nose and continues down the back to the rump. Found in tundra habitats, including rocky hillsides and alpine meadows.

RICHARDSON'S COLLARED LEMMING *Dicrostonyx richardsoni* 115–150mm, 9–15mm, 35–90g

Similar to other lemmings, in winter they are white with long fur that is dark at base, and develop a bifid, or forked "digging" claw. Long hair plus compact build give appearance of slightly elongated powder puff. In summer they molt to reddish to grayish brown, often with a subdued salt-and-pepper appearance because some of the buff hairs are black tipped and some are red tipped. Sides of body and cheeks are reddish orange; sometimes there is a reddish-orange patch behind the ears, and there is usually a reddish band or collar across the chest. The rest of the belly is reddish to buff-gray. Younger animals have a thin black stripe on back from tip of nose to base of tail. Ungava Collared Lemmings tend to be more uniform brownish gray. Occupies tundra habitats including open, dry areas, lichen-heath flats, and rocky or sandy ridges.

NORTHERN COLLARED LEMMING *Dicrostonyx groenlandicus* 110–177mm, 10–20mm, 30–50g

Stocky, wooly, little lemming with seasonally different pelage. In winter, animals are larger, have white pelage, and develop the digging claw common to lemmings. Guard hairs can be twice as long, and underfur twice as thick as in summer. In summer they are smaller, have light grayish-buff to dark gray pelage, with aspects of buff to reddish brown above, distinguishing them from Richardson's Collared Lemming, which are darker reddish brown. Underparts are grayish white. There is a thin black stripe that starts at the tip of the nose and continues down the back to the rump. The subspecies from the Aleutian islands is large, does not turn white or develop winter digging claws, and may be a full species. Occurs only on Arctic tundra, including high, dry, rocky areas in summer.

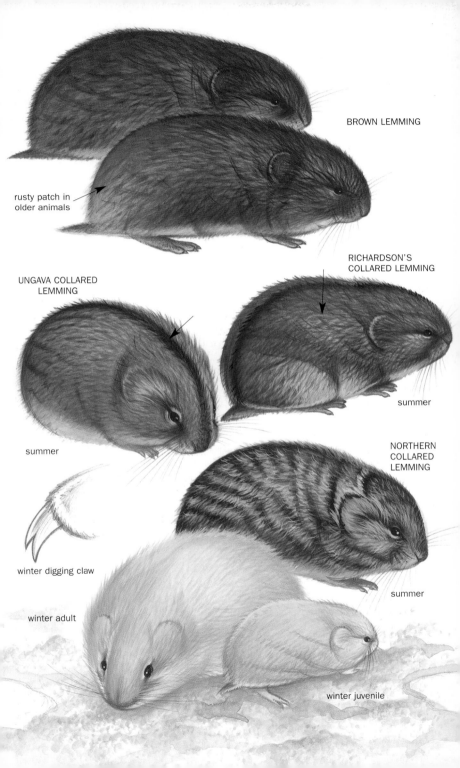

BROWN LEMMING

rusty patch in
older animals

RICHARDSON'S
COLLARED LEMMING

UNGAVA COLLARED
LEMMING

summer

NORTHERN
COLLARED
LEMMING

summer

winter digging claw

winter adult

summer

winter juvenile

PLATE 63
GHOST AND LEAF-NOSED BATS

PETERS'S GHOST-FACED BAT *Mormoops megalophylla* 73–98mm, 20–28mm, 15–16g

Bizarre-looking face with wartlike protuberances on nose and leaflike appendages on chin. Dorsal fur long and lax, with each hair containing four different color zones, the second of which tends to be reddish in mature adults. Forearm 46–56mm. Widely distributed in riparian zones in arid lands up to 3000m.

CALIFORNIA LEAF-NOSED BAT *Macrotus californicus* 85–99mm, 28–41mm, 12–22g

Long ears (more than 25 mm), gray fur, and a distinct, leaflike appendage on the tip of the nose. Tail extends slightly beyond edge of tail membrane. Forearm 46–55mm. May form colonies ranging in size from a few individuals to over 1000. Found in Lower Sonoran lifezones.

MEXICAN LONG-TONGUED BAT *Choeronycteris mexicana* 81–103mm, 6–10mm, 10–25g

Medium-sized leaf-nosed bat with grayish to brownish fur, an elongated muzzle, and a prominent nose leaf. Similar to long-nosed bats, but has a well-developed tail membrane, enclosing a short, conspicuous tail. Forearm 43–49mm. Lacks lower incisors. Forms small colonies in caves, mine tunnels, buildings, and culverts.

MEXICAN LONG-NOSED BAT *Leptonycteris nivalis* 76–88mm, no tail, 18–30g

The only leaf-nosed bat occurring in Texas, this species may overlap with the North American Long-nosed Bat in New Mexico, where it differs by being larger, with grayish fur, a wider tail membrane and shorter wings. Forearm 56–60mm. Emory Peak Cave in Big Bend National Park has housed as many as 10,000 of these bats in mid-summer. The Mexican Long-nosed Bat is classified as an Endangered species; its migratory movements, specialized feeding, and roosting habits make it a conservation risk.

NORTH AMERICAN LONG-NOSED BAT
Leptonycteris yerbabuenae 75–85mm, no tail, 15–25g

Lack of tail distinguishes this species from the California Leaf-nosed and Mexican Long-tongued Bats. Forearm 51–54mm. Summer migrant into southern Arizona and New Mexico, this species occasionally visits hummingbird feeders in search of nectar. The North American Long-nosed Bat is classified as an Endangered species; its migratory movements, specialized feeding, and roosting habits make it a conservation risk.

HAIRY-LEGGED VAMPIRE *Diphylla ecaudata* 67–93mm, no tail, 24–43g

This rare bat has a nose leaf that is greatly reduced and no tail. The incisors are prominent, with very sharp cutting edges. Fur on the back is dark brown, and the belly is slightly paler. Forearm 50–56mm. Normally roosts solitarily in caves or mine tunnels. This species only eats blood, and specializes in feeding on roosting birds. Known in the United States only by a single specimen from Val Verde County, Texas.

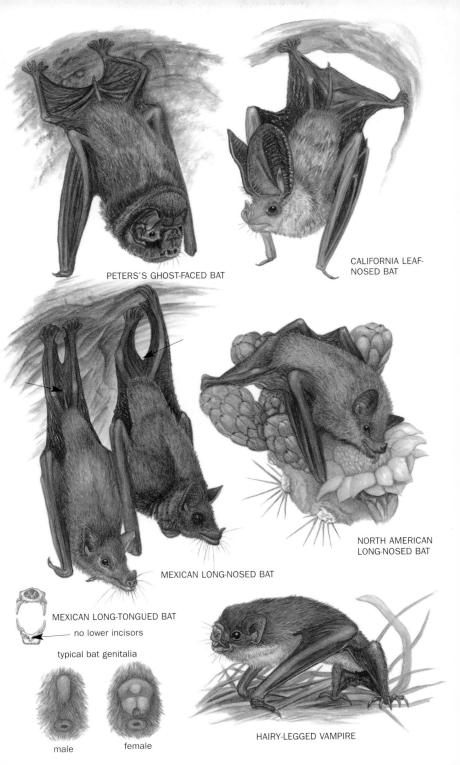

PETERS'S GHOST-FACED BAT

CALIFORNIA LEAF-NOSED BAT

MEXICAN LONG-NOSED BAT

NORTH AMERICAN LONG-NOSED BAT

MEXICAN LONG-TONGUED BAT

no lower incisors

typical bat genitalia

male

female

HAIRY-LEGGED VAMPIRE

PLATE 64
MOLOSSID BATS

FREE-TAILED BATS – This family of bats, the Molossidae, is distinguished by having a tail that extends well beyond the terminal edge of the tail membrane. They emit audible echo-location calls which sound like clicks. Their unique large ears are an adaptation for the echolocation of insect prey in their high-altitude, high-speed hunts.

MEXICAN FREE-TAILED BAT *Tadarida brasiliensis* 85–109mm, 30–39mm, 10–15g

The most common member of the family has deeply furrowed lips, and ears that are not joined at the midline. Forearm 36–46mm. It forms summer nursery colonies of millions of individuals in large caves.

POCKETED FREE-TAILED BAT *Nyctinomops femorosaccus* 99–118mm, 34–45mm, 14–17g

Similar to Mexican Free-tailed Bat, but differs in having ears joined at the base. Forearm 45–50mm. Can be found near large, open-water sources, where it drinks early in the evening.

BIG FREE-TAILED BAT *Nyctinomops macrotis* 120–160mm, 40–57mm, 22–30g

Largest member of the genus. Upperparts range from pale reddish brown to almost black, but individual hairs are white at the base. Large ears are joined at the base, and the lips are wrinkled. Forearm 58–64mm. Inhabits rugged, rocky canyon country in southwestern United States.

GREATER BONNETED BAT *Eumops perotis* 159–187mm, 55–72mm, 45–73g

The largest bat in North America is darker than Underwood's Bonneted Bat, and lacks long guard hairs on the rump. Forearm 72–83mm. Males have a well-developed throat gland that emits a thick, smelly solution in the breeding season. Lives in crevices in vertical cliffs in rugged canyonlands of the arid southwest.

UNDERWOOD'S BONNETED BAT *Eumops underwoodi* 160–165mm, 50–64mm, 40–65g

This large bat has slightly shorter ears than the Greater Bonneted Bat, and has a distinctive fringe of bristle-like guard hairs on the rump. Forearm 65–77mm. They are known only from extreme southern Arizona, where they can be caught drinking at desert pools.

WAGNER'S BONNETED BAT *Eumops glaucinus* ♂ 123–165mm, 40–64mm, 25–47g; ♀ 117–156mm, 40–61mm, 28–55g

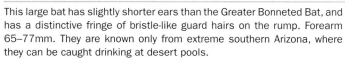

Medium-sized free-tailed bat with smooth lips and large ears that are joined at the base. Forearm 55–68mm. No other similar-sized free-tailed bats occur in its range, which is restricted to southern Florida. Appears to favor old, mature trees as roosting sites, but will use buildings.

PALLAS'S MASTIFF BAT *Molossus molossus* 89–104mm, 30–39mm, 10–14g

Small free-tailed bat with dark brown or grayish-brown fur. Individual hairs are white at their base. Short bristles on rump and wrinkles on the upper lip. Forearm 36–40mm. Tends to roost in attics. Limited to a few colonies roosting in buildings on the Florida keys.

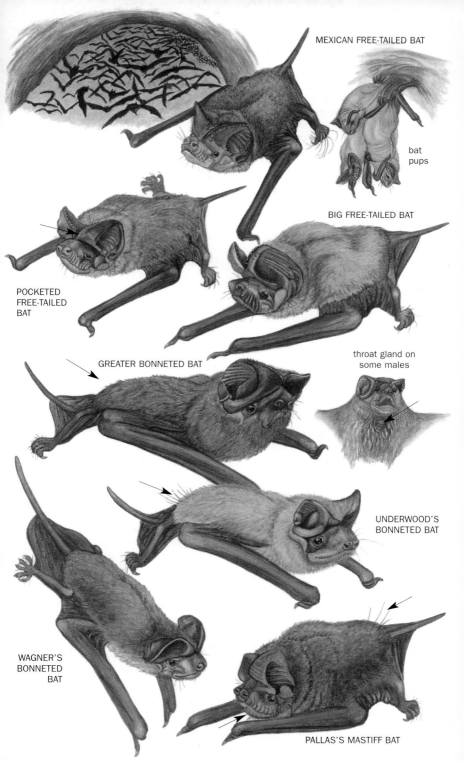

MEXICAN FREE-TAILED BAT

bat
pups

BIG FREE-TAILED BAT

POCKETED
FREE-TAILED
BAT

GREATER BONNETED BAT

throat gland on
some males

UNDERWOOD'S
BONNETED BAT

WAGNER'S
BONNETED
BAT

PALLAS'S MASTIFF BAT

PLATE 65
LASIURINE BATS

LASIURINE BATS – Migratory bats, common throughout the United States in the spring and fall, and widely distributed across North America in the summer. They tend to be solitary, and often roost in trees and foliage.

EASTERN RED BAT *Lasiurus borealis* 95–126mm, 45–62mm, 7–16g

A distinctive looking bat with mottled reddish and grayish pelage that extends well over the interfemoral membrane. The ears are low and rounded; tragus is short and blunt. Red bats have a small projection known as a "lacrimal shelf" at the back of the skull. Forearm 37–42mm (yellow area on map).

WESTERN RED BAT *Lasiurus blossevillii* 92–112mm, 44–52mm, 6–10g

(Not pictured) Recently split from the Eastern Red Bat by genetic data, these two are best distinguished by range (see red area on map of Eastern Red Bat). The Western species differs in being slightly smaller, with fewer frosted hairs on the back and tail membrane.

SEMINOLE BAT *Lasiurus seminolus* 89–114mm, 35–50mm, 9–14g

Similar to female Eastern Red Bats, this species has broadly rounded ears, long pointed wings, and a densely furred interfemoral membrane. Forearm 37–43mm. Favors Spanish moss as a day roost.

NORTHERN YELLOW BAT *Lasiurus intermedius* 121–164mm, 51–77mm, 14–20g

Distinguished by yellowish-gray or yellowish-brown pelage, which extends only onto the basal half of the tail membrane. The ears are pointed. A much larger bat than its relative, the Southern Yellow Bat. Forearm 46–50mm.

SOUTHERN YELLOW BAT *Lasiurus ega* 109–131mm, 42–58mm, 10–14g

Has yellow fur that extends over the basal third to half of the tail membrane, and a short rostrum. Forearm 37–43mm (red area on map).

WESTERN YELLOW BAT *Lasiurus xanthinus* 102–118mm, 38–56mm, 10–15g

(Not pictured) Recently split from the Southern Yellow Bat based on genetic data, the two are best distinguished by range (see yellow area on map of Southern Yellow Bat). The pelage of the Western Yellow Bat is slightly brighter yellow, especially on the tail membrane.

HOARY BAT *Lasiurus cinereus* 99–140mm, 40–64mm, 20–35g

Large, dark bat with grizzled fur that is frosted with white and marked with a yellow collar. The frosting gives this bat a hoary appearance. The fur extends across the tail membrane, and the ears are thickened, short, and rounded. The tragus is short and broad. Forearm 46–55mm.

SILVER-HAIRED BAT *Lasionycteris noctivagans* 90–117mm, 31–50mm, 9–12g

Beautiful dark brown or black bats with frosted hair on the back that extends down onto the basal portion of the tail membrane. The ears are short and rounded, and the tragus is blunt and curved forward. Forearm 37–47mm.

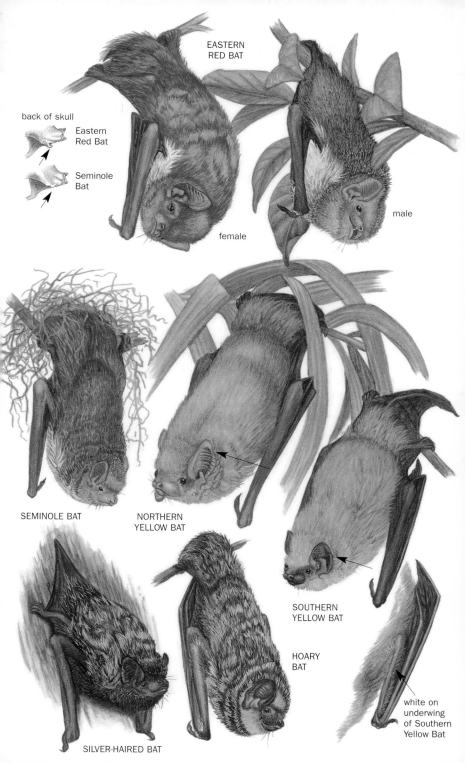

EASTERN
RED BAT

back of skull

Eastern
Red Bat

Seminole
Bat

female

male

SEMINOLE BAT

NORTHERN
YELLOW BAT

SOUTHERN
YELLOW BAT

HOARY
BAT

SILVER-HAIRED BAT

white on
underwing
of Southern
Yellow Bat

PLATE 66
GIANT-EARED BATS

SPOTTED BAT *Euderma maculatum* 107–125mm, 47–55mm, 15–22g

Unmistakable, with three white spots on the back, and large, pinkish ears. One of North America's most beautiful bats. Forearm 48–54mm. It feeds on a variety of insects, particularly moths, and has some echolocation calls that sound like clicks to humans. Widespread but rarely seen, the Spotted Bat lives in the mountain and basin country of western North America.

RAFINESQUE'S BIG-EARED BAT *Corynorhinus rafinesquii* 80–110mm, 42–54mm, 8–14g

Small bat with large ears, toe hairs that extend beyond the tips of the claws, bicolored belly fur, and two large, fleshy lumps on each side of the snout. Forearm 38–44mm. Forms small colonies in a wide variety of roosts, including man-made structures, caves, hollow trees, and under loose tree bark.

TOWNSEND'S BIG-EARED BAT *Corynorhinus townsendii* 89–116mm, 33–54mm, 9–12g

Has huge ears and a pair of glandular lumps on either side of the nose. The back is pale to reddish brown and the belly is pale buff. The western analog of Rafinesque's Big-eared Bat, it is distinguished by belly fur color. Lacks the leaflike lappets of Allen's Big-eared Bat. Forearm 39–47mm. Uncommon and sensitive to disturbance at the roosts. Subspecies from the Ozark Mountains (*C. t. ingens*) and limestone caves of Kentucky, West Virginia and Virginia (*C. t. virginianus*) are endangered. Forages late, which makes it difficult to observe in flight. Often hunts in edge habitats between forest and open areas.

ALLEN'S BIG-EARED BAT *Idionycteris phyllotis* 103–135mm, 40–53mm, 8–16g

Unique with big ears and small lappets that project forward from the base of each ear, extending over the snout. This small bat looks large, owing to huge ears and long, lax, yellowish-gray to blackish-brown fur. Forearm 41–49mm. It is known from a variety of both arid and wooded habitats in southwestern North America.

PALLID BAT *Antrozous pallidus* 92–135mm, 35–53mm, 13–29g

A whitish bat with orange shoulders and large, well-separated pinkish ears. The eyes are relatively large, and the large muzzle is bare. Forearm 48–60mm. This bat has a distinctive odor, which emanates from glands on the muzzle. Unique foraging style allows it to pick up prey, such as scorpions, from the surface of the ground. Uses a variety of arid and semi-arid habitats.

long toe
hairs

SPOTTED
BAT

RAFINESQUE'S
BIG-EARED BAT

TOWNSEND'S
BIG-EARED BAT

ALLEN'S BIG-
EARED BAT

PALLID BAT

PLATE 67
LONGER-EARED *MYOTIS*

MYOTIS BATS – This diverse group of small-bodied bats are among the most common species in North America. They are small, with upperparts generally brownish, and underparts somewhat paler. All are insectivorous, and most roost in colonies, ranging in size from a few dozen to thousands of individuals.

SOUTHWESTERN MYOTIS *Myotis auriculus* 85–101mm, 34–49mm, 6–10g

Medium-sized, with long, brownish ears. Lacks the visible fringe of hairs on the tail membrane that the Fringed Myotis has, and differs from the Long-eared Myotis in having slightly shorter ears (less than 21mm), paler flight membranes, and hair on the back that is brownish rather than blackish at the base. Forearm 38–40mm. Roosts in small groups in caves. Feeds by hovering and gleaning large insects from tree trunks and buildings. Occurs from desert grasslands up into coniferous forests in southwestern mountains.

FRINGED MYOTIS *Myotis thysanodes* 80–99mm, 35–45mm, 6–12g

Easily distinguished by long ears and distinct fringe of hairs along posterior edge of tail membrane. Has the shortest ears among this group of long-eared *Myotis*. Forearm 40–46mm. Females and young found in maternity colonies in caves, mines, and buildings at middle elevations from April to September. Males more likely to roost alone. Occurs in wide range of desert, grassland, and woodland habitats.

LONG-EARED MYOTIS *Myotis evotis* 87–100mm, 36–41mm, 5–8g

Differs from other long-eared *Myotis* in slightly larger body size, longer (more than 21mm) ears that are glossy dark brown to black, and lack of fringe on tail membrane. Forearm 36–41mm. Roosts singly or in groups of up to 30 under bark, bridges, and rocks, and in buildings, caves, crevices, hollow trees, mines, and sink holes. Forages in vegetated areas, where it gleans insects from the surface of the vegetation. Favors coniferous forest, but occurs into riparian desert scrub throughout much of western North America.

KEEN'S MYOTIS *Myotis keenii* 63–93mm, 35–44mm, 4–6g

Small long-eared bat with glossy dark brown fur on back, darker shoulder spots, and ears that extend slightly beyond nose when bent forward. The tail membrane has a few scattered hairs, but no obvious fringe. Forearm 34–39mm. Found individually or in small groups in caves or under the bark of trees. Hibernates colonially, occasionally with other species. Occupying the smallest range of any North American bat, Keen's Myotis lives in dense pacific coastal forests of British Columbia, extending barely into southeastern Alaska and northwestern Washington.

NORTHERN MYOTIS *Myotis septentrionalis* 80–96mm, 29–46mm, 4–11g

A long-eared bat with long, sharply pointed tragus. Differs from Long-eared Myotis in having slightly paler ears, and from Keen's Myotis in having slightly paler shoulder spots and lacking the few scattered hairs on the tail membrane. Forearm 35–40mm. Hibernates in caves and mines. Nursery roosts are under loose tree bark, or in buildings. Prefers coniferous forests, but lives in a variety of woodlands.

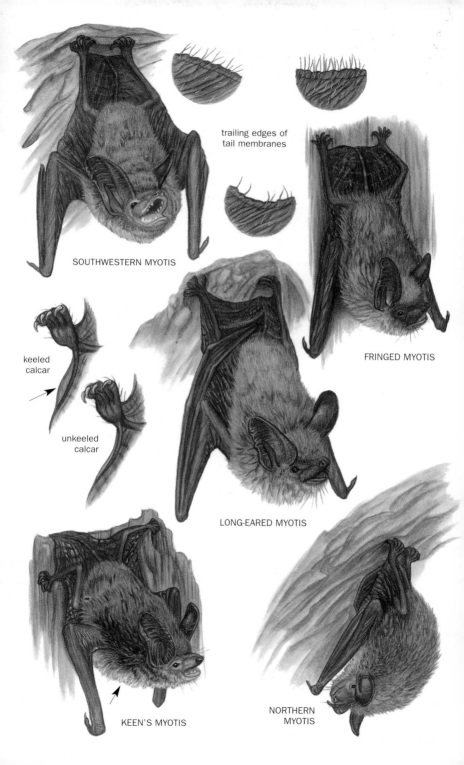

trailing edges of
tail membranes

SOUTHWESTERN MYOTIS

FRINGED MYOTIS

keeled
calcar

unkeeled
calcar

LONG-EARED MYOTIS

KEEN'S MYOTIS

NORTHERN
MYOTIS

PLATE 68
WESTERN *MYOTIS* 1

CALIFORNIA MYOTIS *Myotis californicus* 70–94mm, 31–42mm, 3–5g

Tiny bat with short ears, relatively short hind feet, and an obviously keeled calcar. Differs from Long-legged Myotis in having shorter forearm and tibia, and less fur on ventral surface of wing membrane. Very similar to Western Small-footed Myotis, but California Myotis is slightly more delicate, often has tricolored pelage, and hair that extends further down the forehead. Forearm 30–35mm. Flies slowly but acrobatically early in the evening, often less than 3m above the ground, and frequently over or near water. Occupies deserts and arid interior mountain basins in western North America.

WESTERN SMALL-FOOTED MYOTIS *Myotis ciliolabrum* 76–90mm, 32–44mm, 3–7g

Similar to California Myotis in having short ears and hind feet, and a keeled calcar, but differs from that species in appearing slightly more robust, having bicolored pelage with a slight, glossy sheen, a slightly longer rostrum, and less hair on the forehead. Forearm 30–34mm. In the high plains east of the continental divide, animals are paler in color and occupy rocky outcrops in short grass habitats. West of the divide, animals are larger and more richly colored, and occupy rocky areas in coniferous forests.

LONG-LEGGED MYOTIS *Myotis volans* 76–106mm, 29–49mm, 5–10g

Large-sized *Myotis* with longer, denser fur on the underside of the wing between the knee and elbow than in other species of *Myotis*. Has short, keeled calcar, and short, rounded ears. Color varies from dark brown to reddish buff, with darker ears and membranes. Forearm 35–42mm. Occupies rugged, mountainous terrain, most commonly 2000–3000m. Primarily limited to coniferous forests, but also found in oak and riparian woodlands, extending down into desert areas.

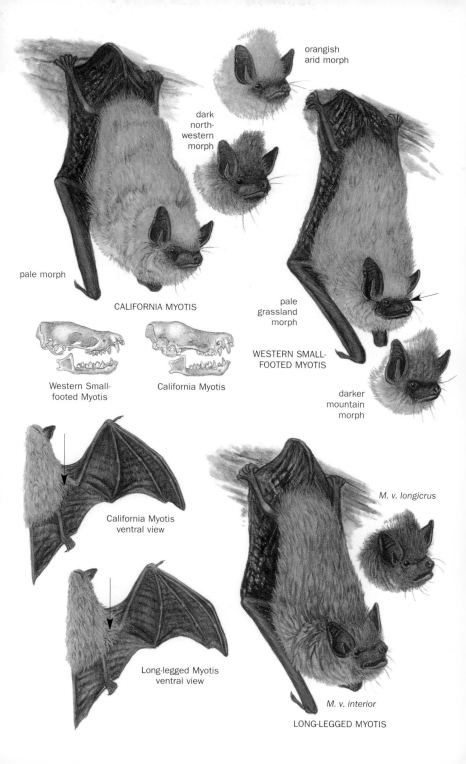

orangish
arid morph

dark
north-
western
morph

pale morph

CALIFORNIA MYOTIS

pale
grassland
morph

WESTERN SMALL-
FOOTED MYOTIS

darker
mountain
morph

Western Small-
footed Myotis

California Myotis

California Myotis
ventral view

Long-legged Myotis
ventral view

M. v. longicrus

M. v. interior

LONG-LEGGED MYOTIS

PLATE 69
WESTERN *MYOTIS* 2

CAVE MYOTIS *Myotis velifer* 83–102mm, 39–47mm, 9–14g

One of the largest species of *Myotis* in North America, with fur varying from light brown to almost black. Has a stubby-nosed appearance, with the ears barely reaching to the end of the nose when bent forward. Forearm 40–45mm. Flies stronger and less erratically when foraging than most other species of *Myotis*. Forms large colonies in caves in lowlands of southwestern North America.

LITTLE BROWN MYOTIS *Myotis lucifugus* 60–102mm, 25–55mm, 7–13g

An otherwise undistinguished species of *Myotis* that differs from the Yuma Myotis in having glossy fur, from the Indiana Bat in lacking a keel on the calcar and having longer toe hairs, from the Long-legged Myotis in having less fur on the underside of the wing, and from Southwestern, Long-eared, Keen's, and Northern Long-eared Myotis in having shorter ears (less than 16mm). Skull shows a more gradually ascending forehead than in the Yuma Myotis. Forearm 33–41mm. Widespread and common in buildings in summer, and hibernates in sizeable colonies in caves and mines.

YUMA MYOTIS *Myotis yumanensis* 75–89mm, 29–43mm, 4–7g

Medium-sized *Myotis* with short ears (extend less than 2mm beyond nose when bent forward), and an unkeeled calcar. Very similar to the Little Brown Myotis, but the Yuma Myotis is slightly smaller, with less glossy fur, and paler ears. Skull shows a more abruptly ascending forehead than the Little Brown Myotis. Forearm 30–38mm. Frequently seen foraging low over or near open water in valleys of western mountains. Commonly found in desert areas, but never far from water sources.

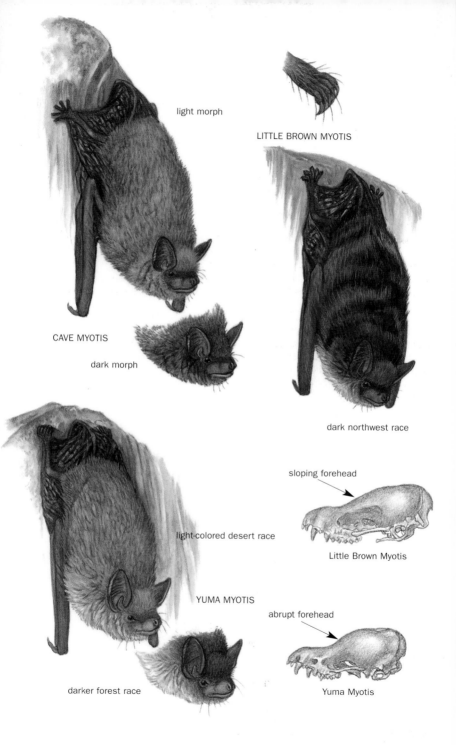

light morph

LITTLE BROWN MYOTIS

CAVE MYOTIS

dark morph

dark northwest race

sloping forehead

Little Brown Myotis

light-colored desert race

YUMA MYOTIS

abrupt forehead

darker forest race

Yuma Myotis

PLATE 70
EASTERN *MYOTIS*

LITTLE BROWN MYOTIS *Myotis lucifugus*

Differs from Indiana Myotis in lacking a keel on the calcar and having more and longer toe hairs. (See page 152 for more details.)

GRAY MYOTIS *Myotis grisescens* 80–96mm, 32–45mm, 4–6g

A large *Myotis* with uniformly-colored gray or brown hairs on the back. Unique in having the wing membranes attached to the ankles rather than to the toes, and the calcar lacks a keel. Forearm 40–46mm. Hind feet are less haired than in the Little Brown Myotis. Females form large maternity colonies in caves in the summer. Endangered, virtually the entire population hibernates in nine caves in the southeastern United States. The Gray Myotis forages mainly over water and ranges over long distances from summer day roosts.

EASTERN SMALL-FOOTED MYOTIS *Myotis leibii* 73–82mm, 29–36mm, 3–7g

The smallest North American *Myotis* differs from other eastern *Myotis* in its smaller size, shorter ears and hind feet, and keeled calcar. Face and ears also darker black than other overlapping bats. Many roosting bats have orange chiggers on their ears. Small feet have no hair on the toes. Similar to Western Small-footed and California Myotis, but range does not overlap with those species. Individual hairs are varying shades of brown, but overlain with a somewhat golden sheen. Forearm 30–35mm. Uncommon and limited to eastern deciduous and coniferous forests.

SOUTHEASTERN MYOTIS *Myotis austroriparius* ♂ 77–89mm, 26–44mm, 5–7g; ♀ 80–97mm, 29–42mm, 5–8g

Small bat with somewhat dull, wooly pelage. Usually dull gray to gray-brown, but some individuals are a bright orange-brown. Fur on underside is dark brown to black at the base, with contrasting white at tip. No prominent keel on calcar. Toes have long hairs extending beyond tips of claws. Differs from Northern Myotis by having smaller ears, and from Little Brown Myotis in lacking the glossy pelage of that species. Forearm 33–42mm. Prefers caves near water, or even those that contain water. Can be seen foraging near the surface of the water on summer evenings.

INDIANA MYOTIS *Myotis sodalis* 73–99mm, 29–43mm, 3–10g

Small bat most closely resembling the Little Brown Myotis. Differs from that species in having sparser and shorter hairs on the toes, shorter hind feet, a slight keel on the calcar, a pinkish nose and duller pelage. Forearm 31–40mm. Hibernates in caves and mines, where it sometimes forms large colonies. This Endangered species is sensitive to disturbance in the hibernacula. Summer nursery roosts are under tree bark.

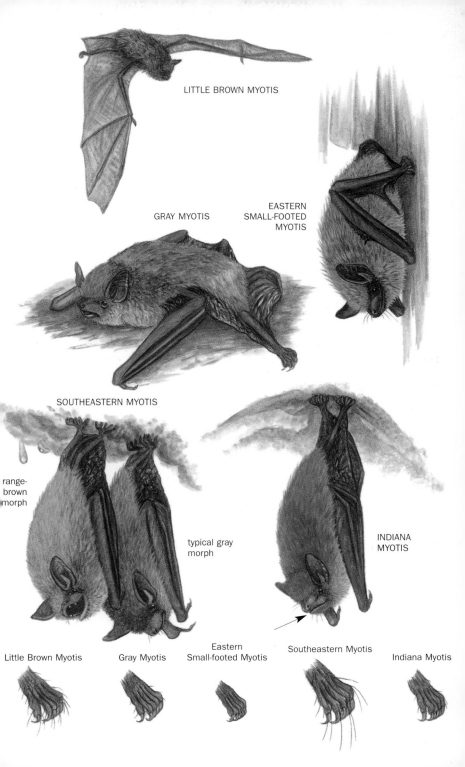

LITTLE BROWN MYOTIS

EASTERN
SMALL-FOOTED
MYOTIS

GRAY MYOTIS

SOUTHEASTERN MYOTIS

range-
brown
morph

typical gray
morph

INDIANA
MYOTIS

Little Brown Myotis   Gray Myotis   Eastern
Small-footed Myotis   Southeastern Myotis   Indiana Myotis

PLATE 71
OTHER EASTERN BATS

WESTERN PIPISTRELLE *Pipistrellus hesperus* 60–86mm, 25–36mm, 2–6g

Tiny bat with blunt, slightly curved tragus, and small hind foot less than half as long as the tibia. The black face mask contrasts strongly with yellowish, gray-brown body color. Active in the early twilight, this species has slow erratic flight that looks almost like a butterfly. Forearm 26–33mm. Found from sea level to 3000m, favors canyons and cliff faces, but occurs in a wide variety of desert scrub and arid grassland habitats as well.

EASTERN PIPISTRELLE *Pipistrellus subflavus* 75–90mm, 33–45mm, 6–10g

Small bat with yellowish- to grayish-brown pelage, and a somewhat paler belly. Dorsal hairs are tricolored, and membranes are dark brown. In winter, hibernating individuals sometimes collect droplets of moisture on them, given them a sparkling appearance. Forearm 31–35mm. Hangs with a hunched posture while sleeping. Individuals swarm in late summer and early autumn, seeking mates before entering hibernating caves. Uses a variety of habitats, especially open pasture.

BIG BROWN BAT *Eptesicus fuscus* 87–138mm, 34–56mm, 11–23g

As the name suggests, this is a large brown bat with naked, dark membranes. The tail is enclosed in interfemoral membrane. It has short ears, a keeled calcar, and a forearm length of 39–54mm. Although hibernacula in caves are common, this species also sometimes hibernates in cold buildings. Widely distributed throughout North America, this species occurs in a wide variety of habitats, and frequently roosts in man-made structures.

LITTLE BROWN MYOTIS *Myotis lucifugus*

Obviously browner than the pipistrelles, smaller than the Big Brown Bat, and not white on the belly like the Evening Bat. (See page 152 for more details.)

EVENING BAT *Nycticeius humeralis* 83–96mm, 35–40mm, 9–14g

Similar to the Big Brown Bat, but smaller, with reddish to dark brown fur above and somewhat paler below. The ears are blunt and rounded, with a tragus that is short and curves forward. Forearm 34–37mm. One of the few bats that does not use caves, the Evening Bat apparently migrates southward to overwinter, rather than hibernating. Common in buildings and other man-made structures.

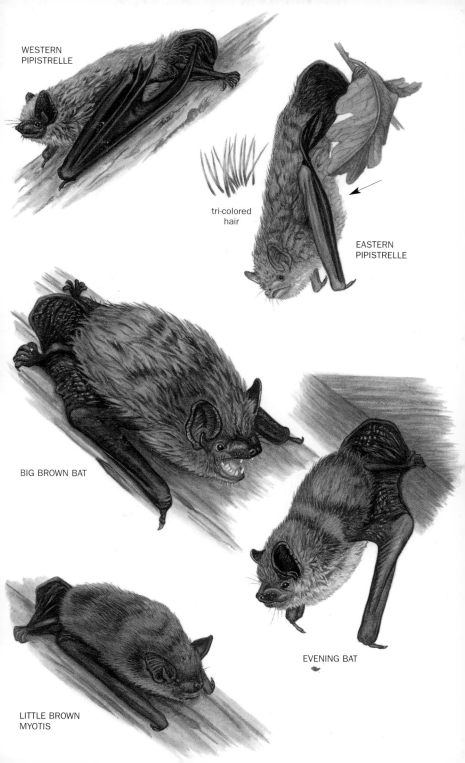

WESTERN
PIPISTRELLE

tri-colored
hair

EASTERN
PIPISTRELLE

BIG BROWN BAT

EVENING BAT

LITTLE BROWN
MYOTIS

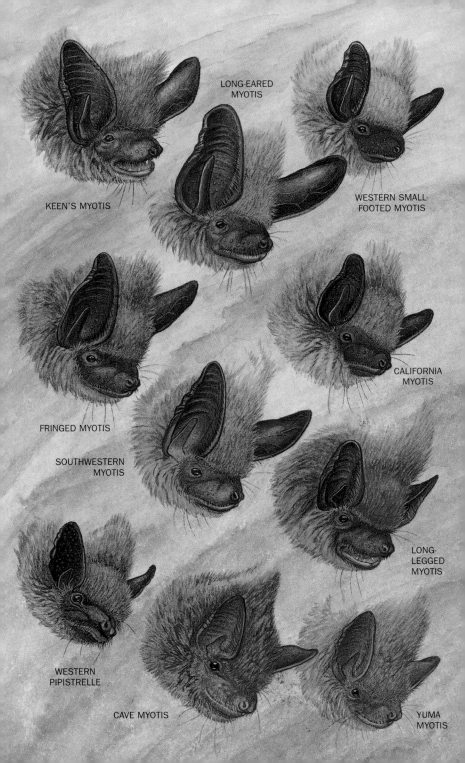

LONG-EARED
MYOTIS

KEEN'S MYOTIS

WESTERN SMALL-
FOOTED MYOTIS

CALIFORNIA
MYOTIS

FRINGED MYOTIS

SOUTHWESTERN
MYOTIS

LONG-
LEGGED
MYOTIS

WESTERN
PIPISTRELLE

CAVE MYOTIS

YUMA
MYOTIS

PLATES 72 & 73
**CONFUSING
CAVE
CHIROPTERA**

EASTERN
SMALL-FOOTED MYOTIS

NORTHERN
MYOTIS

INDIANA MYOTIS

LITTLE BROWN MYOTIS

BIG BROWN BAT

EASTERN
PIPISTRELLE

SOUTHEASTERN MYOTIS

GRAY MYOTIS

PLATE 74
TEMPERATE CATS

BOBCAT *Lynx rufus* ♂ 48–125cm, 11–20cm, 7.2–31.0kg; ♀ 61–122cm, 9–17cm, 4–24kg

North America's most abundant spotted cat, with a tail tip that is black above and white below. Tail is relatively longer than in the Canadian Lynx. Coat color varies from yellowish to reddish brown and is marked with dark brown or black streaks and spots. Spots may be inconspicuous. Underparts are white and spotted. Ears have short (less than 2.5cm) black tufts and a white spot on the back. Primarily a nocturnal hunter of rabbits, and other small prey. Occurs in almost every habitat within range.

CANADIAN LYNX *Lynx canadensis* 82–95cm, 9–12cm, 7–18kg

This northern species has a tail that is all black at the tip. Upperparts are grizzled grayish brown with a varying degree of spotting. Winter fur is long and thick. Ears have long (4–5cm) black tufts of hairs and a central white spot on the back. Tail is relatively shorter than in the Bobcat; ear tufts and feet are larger. Females are slightly smaller than males. Threatened because of sensitivity to human disturbance. Specialist hunter of Snowshoe Hares; also hunts other small mammals, beaver and deer. Denizen of boreal forests.

PUMA *Puma concolor* ♂ 1.7–2.5m, 68–96cm, 36–120kg; ♀ 1.2–1.6m, 27–37cm, 29–64kg

An unmistakable large cat with uniform color and long black-tipped tail. Monotone pelage varies geographically and seasonally from gray to reddish brown. Cubs have a soft, spotted coat. Hunts deer and other large and medium-sized mammals. Subspecies in Florida (*P. c. coryi*) is Endangered. Subspecies from northeastern United States (*P. c. couguar*) is apparently extinct; modern sightings in this area are from escaped captive animals. Uses most habitats in range that offer cover and prey; avoids shrubless deserts and agricultural areas.

BOBCAT

CANADIAN LYNX

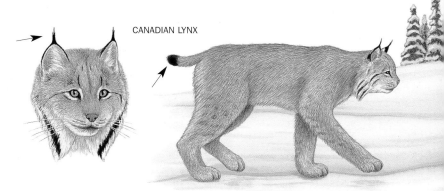

PUMA

cub

PLATE 75
TROPICAL CATS

TROPICAL CATS – These four felines range widely in tropical America, but are only found near the borderlands of the United States. Because of their secretive nature they are rarely seen.

MARGAY *Leopardus wiedii* ♂ 86–130cm, 33–51cm, 3–7kg; 80–103cm, 32–44cm, 3–5kg

Smallest North American spotted cat, with large eyes and a long tail. Smaller than the Ocelot, with a longer tail, relatively larger eyes, and a more delicate facial structure. Spots may coalesce to a greater or lesser extent. A hunter of small prey on the ground and in the trees. Probably never common in Texas, the last record was from 1852. Uses most tropical forest types.

OCELOT *Leopardus pardalis* ♂ 95–137cm, 28–40cm, 7–14kg;
♀ 92–121cm, 27–37cm, 7–11kg

A solidly built spotted cat, with a short tail that barely touches the ground. Larger than the Margay, with a shorter tail and a more robust skull structure. Rare and Endangered in Texas. This nocturnal hunter feeds on a variety of small prey, especially rodents. Uses a wide range of tropical habitats with vegetative cover.

JAGUARUNDI *Puma yagouaroundi* 88–137cm, 32–61cm, 4–9kg

Only small wildcat with a unicolored coat. Two color morphs occur together: reddish brown and dark gray. Young animals may have white markings around their mouth. Tail and neck are relatively long, legs are relatively short. Our most diurnal cat; hunts a variety of small, terrestrial prey. Endangered in North America because of small range. Uses thick cover in mesquite or pine-oak woodlands of Arizona and Texas.

JAGUAR *Panthera onca* ♂ 1.7–2.4m, 52–67cm, 79–158kg; ♀ 1.6–2.2m, 43–60cm, 31–85kg

An unmistakable, powerfully built, spotted cat. Coat is tawny-colored with large black rosettes. Rosettes are also visible in the rare black morph. Active any time of day. The US population was hunted to extinction in the 1800s, but recent sightings in Arizona suggest the possibility of reestablishment. Hunts large and medium-sized prey in warm forests and desert scrub.

MARGAY

OCELOT

JAGUARUNDI

dark
phase

light phase

JAGUAR

PLATE 76
CANIS

CANIS – All North American species in this genus are capable of interbreeding in the wild. This causes confusion for both identification and taxonomy. The major forms are described here, although their taxonomic status and nomenclature may change in the future. These pack-living canids are predominantly nocturnal, but may be active any time.

GRAY WOLF *Canis lupus* ♂ 1.0–1.3m, 35–52cm, 30–80kg; ♀ 0.9–1.2m, 35–52cm, 23–55kg

Largest wild dog. Fur usually grizzled gray but ranges from pure white to black to reddish brown. Bushy tail often tipped in black. Distinguished from Coyote and domestic dogs by larger body size, large nose pad (>25mm), relatively shorter ears, and tail usually held horizontally (not down) when running. Canids between the Wolf and Coyote in size (typically 14–39kg) remain a taxonomic problem. Some taxonomists argue that all intermediates are recent hybrids between Wolves and Coyotes and therefore do not merit species' status. Others argue they are not hybrids, or that they are ancient hybrids, and that they deserve full status as one or two distinct mid-size *Canis* species: the Eastern Canadian Wolf (*C. lycaon*) from the north and/or the Red Wolf (*C. rufus*) from the southeast. Packs usually consist of 5–10 family members. Howling can be heard for great distances. Common where not persecuted; lives in all habitats within its range except deserts and high mountain tops.

COYOTE *Canis latrans* 1.0–1.3m, 30–39cm, 7–20kg

Smaller than wolves, with a smaller nose pad (less than 25mm) but relatively larger ears. Upperparts are typically brownish, but variable; mountain animals may be gray or black; desert animals are reddish brown. Belly and throat are pale. Tail usually held down when running. Expanded range east in mid-1900s following extirpation of wolves; eastern animals larger; northern animals have longer and coarser fur. Females average slightly smaller. Larger packs where not persecuted. Higher-pitched howling than wolves, with more yips. Common in all natural habitats, including suburban forests.

DOMESTIC DOG *Canis familiaris*

Variable, but rarely show characteristic ears and face of wild canids. Coydogs (coyote × dog hybrids) only occur with very low coyote density, practically nonexistent now that coyotes have populated most of the continent. Packs of feral Domestic Dogs persist, especially in the south (e.g., Florida, Texas).

black morph howling

GRAY WOLF

white morph

"Red Wolf"

"Eastern Canadian Wolf"

eastern

COYOTE

ground den

western

DOMESTIC DOG

PLATE 77
SMALLER FOXES

FOXES – These diminutive canids typically hunt small and medium-sized prey such as birds, rabbits, and mice. All den in ground holes or brush piles.

SWIFT FOX *Vulpes velox* ♂ 74–82cm, 17–33cm, 2.2–2.9kg; ♀ 68–75cm, 23–30cm, 1.8–2.3kg

Smallest North American canid, with a long, bushy, black-tipped tail. Buff-gray above with brownish sides, legs and lower surface of tail. Undersides are lighter. Shorter summer fur is more reddish gray. Large triangular ears. Often lumped with the Kit Fox into one species (*V. velox*). The Kit Fox is distinguished by its longer tail (62% head and body length vs. 52%), larger ears (more than 75mm) set close to the midline of the skull, a head that is broader between the eyes but narrower at the snout, and more slitlike, less-rounded eyes. Nocturnal hunter of lagomorphs, rodents, and other small prey. The two species also differ in habitat preference, with the Swift Fox preferring short and mixed grass prairies.

KIT FOX *Vulpes macrotis* 73–84cm, 26–32cm, 1.4–2.7kg

Similar to the Swift Fox but with a longer tail and larger ears. See Swift Fox account for details. Feeds predominantly on rodents and lagomorphs. Prefers shrub-steppe and desert habitats.

ARCTIC FOX *Alopex lagopus* ♂ 83–110cm, 28–42cm, 3.2–9.4kg; ♀ 71–85cm, 26–32cm, 1.4–3.3kg

Unique northern fox with distinctive color and short legs, snout, and ears. Color of long fur varies geographically and seasonally. Coastal populations are typically bluish gray in winter and dark bluish gray or chocolate-brown in summer. Continental foxes are usually white in winter and grayish brown in summer. Eyes have golden irises; soles of feet are furred. Often bold in the presence of humans. Nocturnal except in Arctic summer. Hunts seabirds and small mammals alone, may scavenge in pairs or small groups. Inhabits Arctic tundra.

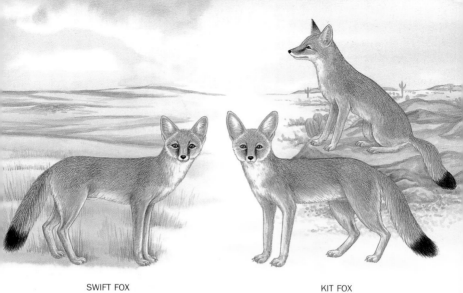

SWIFT FOX

KIT FOX

inland winter

inland summer

ARCTIC FOX

coastal summer

coastal winter

PLATE 78
LARGER FOXES

RED FOX *Vulpes vulpes* ♂ 74–82cm, 27–33cm; ♀ 68–75cm, 23–30cm, 3–6kg

Uniquely-colored fox with a bushy, white-tipped tail. Most animals are red-colored with black feet and black-tipped, triangular ears. Other morphs include a black phase, a silver phase (black hair tipped with silver), and a cross phase (reddish brown with dark on shoulders). Recently declining in the east due to Coyote expansion. Prefers landscapes with a mixture of both open fields and brushy or forested areas.

GRAY FOX *Urocyon cinereoargenteus* 80–113cm, 27–43cm, 3–7kg

A gray fox with a black-tipped tail. Back is gray; sides are cinnamon-colored; belly is tan. A dark stripe runs down center of back and onto black-tipped tail. Often climbs trees. Our most omnivorous canid, eats fruit and prey the size of rabbits or smaller. Active mostly at dawn and dusk. Uses hardwood forests and brushy riparian habitats.

ISLAND GRAY FOX *Urocyon littoralis* ♂ 62–79cm, 14–29cm, 1.6–2.5kg; ♀ 59–79cm, 11–29cm, 1.5–2.3kg

Only found on six Channel Islands off California coast. Like Gray Fox, but smaller (one half to two thirds in size). Active day and night. Remarkably docile. Feeds on fruit, insects, and small mammals. A conservation risk because of small range; feral pigs and goats on the islands may attract avian predators (Golden Eagles) to which this isolated carnivore has no defense. Lives in all habitats on islands.

typical

RED FOX

"Silver Fox"

"Cross Fox"

Red Fox
pups

GRAY FOX

ISLAND GRAY
FOX

PLATE 79
BEARS

BLACK BEAR *Ursus americanus* ♂ 1.5–2.1m, 8–14cm, 47–409kg; ♀ 1.3–1.7m, 8–14cm, 39–236kg

Distinguished from other bears by smaller size, larger ears, pale muzzle, and a rounded back. Fur color varies geographically; most eastern animals are dark black; western populations can be brown, cinnamon, or blond. Some coastal populations in British Columbia and Alaska are creamy white (Kermode Bears) or bluish gray (Glacier Bears). Some animals have a white chest patch. Lips are prehensile. Males are larger. Often leaves its mark on trees when stripping bark to eat sap, climbing tree with claws, or rubbing and scratching to mark territory. In most areas the Black Bear hibernates through the winter in ground or tree dens; in the south only pregnant females hibernate. Populations are increasing across most of their range. Omnivorous, it is an opportunistic predator in woodlands and swamps.

GRIZZLY BEAR *Ursus arctos* 1.0–2.8m, 6–21cm, 80–600kg

Large brownish bear with a dished facial profile and a humped shoulder. Fur color variable and may be virtually any shade of brown. Head and shoulders are typically paler than the darker sides, belly, and legs. Front claws are extremely long. Variable in size with larger coastal (Brown Bear) and island (including Kodiak Bear) populations and smaller inland (Grizzly Bear) forms. Males are larger. Leaves marks on trees like Black Bears. Omnivore and predator. Threatened and declining through much of its range in North America and Europe. Seasonally abundant near salmon spawning streams. Persists in remote forests, tundra, and open plains.

POLAR BEAR *Ursus maritimus* ♂ 2.3–2.6m, 7–12cm, 400–800kg; ♀ 1.9–2.1m, 7–12cm, 175–300kg

A very large, white bear of the Arctic. Longer neck and relatively smaller head than other bears. Fur may appear yellow in summer. Pregnant females may weigh up to 500kg. Threatened by melting ice associated with global warming and airborne pollutants that accumulate in polar regions. Pursues fish and seal prey in pack ice and coastal regions.

"Glacier Bear"

"Kermode Bear"

blond morph

cinnamon morph

eastern black morph

BLACK BEAR

western brown morph

GRIZZLY BEAR

POLAR BEAR

PLATE 80
PROCYONIDS

RINGTAIL *Bassariscus astutus* 62–81cm, 31–44cm, 870–1143g

Unique with slender build and long, bushy, striped tail. Face is gray, back is pale yellowish to tawny reddish in color and the underparts and feet are pale. Parallel-sided tail has 14–16 stripes and a black tip. Males are slightly larger. Strictly nocturnal. Eats fruit, mice, and other small prey. Inhabits woodlands, riparian areas, and arid scrubland.

NORTHERN RACCOON *Procyon lotor* ♂ 63–95cm, 20–40cm;
♀ 60–91cm, 19–34cm, 4.0–15.8kg

Well known for its dark mask and ringed tail. Grizzled pelage varies from gray to blackish. In winter a yellowish or reddish tinge may develop on the nape of neck. Albino, dark brown, and cinnamon color phases are also known. Omnivore and semiaquatic forager. Larger in north, where well-fed animals can reach 50% body fat. Common in every habitat with water sources.

WHITE-NOSED COATI *Nasua narica* 85–164cm, 25–38cm, 2.7–6.5kg

This tropical procyonid reaches the northern edge of its range at the US–Mexico border. Unique with a long mobile snout and a slender tail with incomplete dark rings. Males are larger than females. The Coati is a diurnal forager of invertebrates and fruit. Females and young travel together in bands of up to 40 animals while males are typically solitary. In the United States it can be found in montane and riparian woodlands.

RINGTAIL

NORTHERN
RACCOON

male

WHITE-NOSED COATI

females with young

PLATE 81
AQUATIC MUSTELIDS

AMERICAN MINK *Mustela vison* ♂ 55–70cm, 19–22cm, 550–1250g;
♀ 47–60cm, 15–19cm, 550–1000g

Dark, semi-aquatic weasel with a white chin patch. Back and belly rich brown; variable amount of white on chin and throat. Hind toes slightly webbed. The Sea Mink subspecies (*M. v. macrodon*) was larger and redder, with a strong scent. It used rocky sea shores along our North Atlantic coast until it was hunted and trapped to extinction by the 1860s; some consider it a full species. Solitary and most active at dusk and dawn. Predator of aquatic birds, mammals, fish, and frogs. The American Mink is common, but rarely seen, near the shoreline of waterways and lakes.

NORTHERN RIVER OTTER *Lontra canadensis* ♂ 1.1m, 42–47cm,
7.7–9.4kg; ♀ 0.9–1.1m, 31–40cm, 4.5–13.6kg

The only river otter in North America. Back is brown; chin and throat are silvery. Tail is long, thick at the base, and gradually tapering. Mostly nocturnal, often seen at dusk or dawn. Scenting latrine areas on shore consist of matted-down vegetation and scraped-up earth, with defecations around the periphery. Eats fish and other aquatic prey. Lives in most types of unpolluted freshwater and coastal marine habitats.

SEA OTTER *Enhydra lutris* ♂ 1.3–1.4m, 36cm, 18.0–45.0kg;
♀ 1.1–1.4m, 27cm, 11.0–33.0kg

The only marine mammal with plush fur, pawlike hands, and flipper-like feet. Tail is flattened dorso-ventrally. Noses of females are often scarred from aggressive males. Forages solitarily (females with their pups), but often rests and socializes in groups called "rafts." Rafts in south are small (fewer than 12) but hundreds of males may congregate in the north. A Sea Otter typically floats on its back and handles its invertebrate prey on its belly. Recovering from massive hunting at turn of century, still threatened. Lives in shallow coastal waters.

AMERICAN MINK

Sea Mink (extinct)

latrine

NORTHERN
RIVER OTTER

female

juvenile

SEA OTTER

male

PLATE 82
BIG MUSTELIDS

AMERICAN MARTEN *Martes americana* ♂ 560–680cm, 20–23cm, 470–1250g; ♀ 50–60cm, 18–20cm, 280–850g

Identifiable with bushy tail, buff-orange throat patch, and sharp facial profile. Fur varies from light to dark brown. Legs are short and ears are small and rounded. Larger than American Mink and smaller than Fisher. Predominantly a predator of small mammals and birds, will also eat fruits and insects. Active day and night, but secretive and rarely seen. Sensitive to trapping pressure and logging; recent natural expansions and managed reintroductions are helping Martens reclaim some of their historic range. Arboreal and terrestrial hunter of coniferous forests.

FISHER *Martes pennanti* ♂ 90–120cm, 37–41cm, 3.5–5.5kg; ♀ 75–95cm, 31–36cm, 2.0–2.5kg

Large, dark-colored, stocky weasel. Bigger and darker than American Marten. Face, neck and shoulders sometimes marked with hoary gold or silver-colored guard hairs. Deep brown coat lightens over summer, darkens in fall molt, and may appear reddish in spring. Throat and chest are marked with white or cream patches of varying size and shape. Females are smaller, with finer and glossier fur. Arboreal and terrestrial predator of small and medium-sized animals. Regularly preys on Porcupines with fast frontal attacks; also eats fruits and nuts. Increasing in the east, declining in the west. Hunts in a variety of forest types.

WOLVERINE *Gulo gulo* ♂ 94–107cm, 22–26cm, 11.3–16.2kg; ♀ 86–93cm, 21–25cm, 6.6–14.8kg

Largest member of the weasel family, with a pair of yellowish bands running from shoulder to rump. Has a large head and a short, stout neck. Typically holds tail and head low while walking with an arched back. Wide-ranging scavengers and predators of Caribou and smaller prey. Lives at low densities in tundra and forested habitats.

AMERICAN BADGER *Taxidea taxus* ♂ 60–87cm, 10–15cm, 3.6–11.4kg; ♀ 52–79cm, 10–15cm, 3.6–11.4kg

Unique with short legs and black and white striped face. White stripe runs from nose to neck in most populations; stripe continues to base of tail in the southwestern animals. Long coarse fur is gray on the back and may be mixed with white, brown, buff, rust, or orange color. Young are pale buff-colored. An efficient digger with very large claws. Sleeps through most of winter, but becomes active on warmer days. Hunts ground squirrels and other small and medium-sized prey. Lives in grasslands, deserts, and open marshy areas.

AMERICAN
MARTEN

FISHER

WOLVERINE

southwest

AMERICAN BADGER

typical

PLATE 83
MUSTELA

LEAST WEASEL *Mustela nivalis* ♂ 180–205mm, 25–40mm, 40–55g; ♀ 165–180mm, 22–30mm, 30–50g

North America's smallest carnivore; lacks a black tail tip. In summer, back varies from rusty brown to pale sandy tan; belly is white or yellowish. Northern populations have a pure white winter coat. Molting animals may appear spotted. Preys on mice day and night; must eat its body weight in food each day. Prefers marshes, grasslands, and other nonforested areas.

ERMINE *Mustela erminea* ♂ 219–343mm, 65–90mm, 67–116g; ♀ 190–292mm, 42–70mm, 25–80g

A short-tailed weasel with a dark tail tip and (typically) white on feet. Smaller than Long-tailed Weasel, with a relatively shorter tail (less than 44% head-body length). Males are 40–80% larger; animals can also be sexed by genital morphology. Back is brown in summer. Winter coat is white, with a black tail tip. Primarily a nocturnal hunter of mice, young rabbits, and other small prey. Uses most habitats in range.

LONG-TAILED WEASEL *Mustela frenata* ♂ 330–420mm, 132–294mm, 160–450g; ♀ 280–350mm, 112–245mm, 80–250g

A long-tailed weasel with a dark tail tip. Larger than the Ermine, with a relatively longer tail (more than 44% head-body length). Variable amount of white on hind legs and feet. Northern populations molt to a white winter coat. Some populations in Florida and the southwest have white or yellow facial markings (leading to the name Bridled Weasel). Hunts small prey, especially mice and young rabbits. Found in most habitats through large range, especially near water.

BLACK-FOOTED FERRET *Mustela nigripes* ♂ 490–600mm, 107–140mm, 915–1125g; ♀ 479–518mm, 120–141mm, 645–850g

This masked weasel survives only in captivity and in a few recently reintroduced wild populations in western states. Unique with a sandy-colored body and dark feet, tail tip, and facial mask. Nocturnal and terrestrial. In digging for prey it often leaves a distinctive pile of soil on the surface measuring about 100×40×10cm. Specialist hunter of prairie dog towns and surrounding areas in western grasslands.

LEAST WEASEL

winter

summer

female

male

ERMINE

winter

LONG-TAILED
WEASEL

summer

"Bridled Weasel"

BLACK-FOOTED FERRET

PLATE 84
SPOTTED AND HOG-NOSED SKUNKS

SPOTTED SKUNKS – Unique with their glossy, jet-black fur and white spots beginning near the ears and forming a pair of dashed lines down the back. Although some consider them to be one species, these two small skunks were recently split into an Eastern and Western species based on genetic and reproductive evidence: the Western Spotted Skunk has delayed implantation but the Eastern does not. Externally the Western Spotted Skunk has more extensive white markings than its Eastern cousin. They typically warn potential predators with their trademark handstand before spraying. These two quickest and most agile skunks are also the most carnivorous, catching mice, insects, and other small prey in their nocturnal hunts.

WESTERN SPOTTED SKUNK *Spilogale gracilis* ♂ 35–58cm, 10–21cm, 500–900g; ♀ 32–47cm, 200–600g

A small spotted skunk with large white spots on the face, back, and tail. Uses rocky and brushy areas in woodlands, grasslands, farmlands, and deserts.

EASTERN SPOTTED SKUNK *Spilogale putorius* ♂ 31–61cm, 8–28cm, 276–885g; ♀ 27–54cm, 9–21cm, 207–475g

Small skunk with thin white spots and a small white tip to a mostly black tail. Larger than Western Spotted Skunk in area of overlap. Avoids open areas, preferring habitats with extensive cover, especially riparian woodlands.

HOG-NOSED SKUNKS – Named for their unique long, naked nose pads, these skunks have coarse, long black fur marked with a white stripe down the back. The extent of this stripe on the rump, and color of the tail underside distinguishes the Eastern (narrow stripe, black under base of tail) and Western (wide stripe, predominantly white tail) species. Recent genetic evidence suggests lumping these two forms into one species. Claws are well developed for digging up insect prey. Nocturnal.

EASTERN HOG-NOSED SKUNK *Conepatus leuconotus* ♂ 56–92cm, 22–41cm, 3.0–4.5kg; ♀ 58–74cm, 19–34cm, 2.0–4.0kg

Found only in southern Texas and Mexico. White dorsal stripe is narrow, and reduced or absent on the rump. Underside of tail is black at the base and white at the tip. Larger than Western Hog-nosed Skunk. Infrequent recent sightings raise concern over conservation status. Recorded from brush habitat and semiopen grasslands.

WESTERN HOG-NOSED SKUNK *Conepatus mesoleucus* 40–84cm, 13–35cm, 1.1–2.7kg

Dorsal stripe is wide, and tail is predominantly white above and below. Males are larger than females, and occasionally reach 4.5kg. Inhabits rocky terrain and stream beds in desert-scrub and mesquite grasslands.

WESTERN SPOTTED SKUNK

EASTERN
SPOTTED
SKUNK

EASTERN HOG-NOSED SKUNK

WESTERN
HOG-NOSED SKUNK

PLATE 85
MEPHITIS

STRIPED SKUNK *Mephitis mephitis* 57–80cm, 18–39cm, 1.2–6.3kg

Dorsal stripes converge to a V at the nape. A pair of dorsal stripes typically mark the back, but these may be so variable in size and shape that skunks look all white, all black, or spotted. The amount of white in the tail is similarly variable. Fur is coarse. Females are 15% smaller than males. Typically raises tail and stomps front feet before spraying. Nocturnal hunter of insects and small vertebrates in woodlands, fields, agricultural areas, and human neighborhoods.

HOODED SKUNK *Mephitis macroura* ♂ 56–79cm, 27–43cm, 800–900g; ♀ 65cm, 37cm, 400–700g

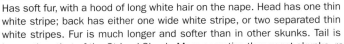

Has soft fur, with a hood of long white hair on the nape. Head has one thin white stripe; back has either one wide white stripe, or two separated thin white stripes. Fur is much longer and softer than in other skunks. Tail is longer than that of the Striped Skunk. More secretive than most skunks, rarely dens in man-made structures. Nocturnal forager on insects and other small prey. Prefers rugged, rocky canyons or heavily vegetated stream edges in lowland desert areas (below 2500m elevation).

typical morph

STRIPED SKUNK

"spotted" Striped Skunk

HOODED SKUNK

double, thin-striped morph

single, wide-striped morph

PLATE 86
OTARIID SEALS

OTARIIDS – THE EARED SEALS – These Pacific seals have small, but distinct ears. Their hind flippers can rotate under their body, increasing their mobility on land.

CALIFORNIA SEA LION *Zalophus californianus* ♂ 2.0–2.5m, 350–400kg; ♀ 1.6–1.8m, 90–120kg

Unique with doglike face and brown color. Males are larger, darker, and have a thick neck and an enlarged, pale forehead. Pups are born dark, but quickly molt into a lighter blond pelage like females. First toe on hind flipper is largest. Males wander north in the winter as far as Vancouver. Often unwary near cities. Feeds on small fish and squid at sea; returns to shore for pupping and mating.

NORTHERN FUR SEAL *Callorhinus ursinus* ♂ 1.9–2.1m, 175–275kg; ♀ 1.2–1.5m, 30–50kg

Head is short and nose is very sharply pointed. Also distinguished from Guadalupe Fur Seal by having fur that forms a straight line at the base of the foreflipper, rather than extending to a point. Females are brown; males grow darker as they age, some appear black, especially when wet. Pups are born black, molt to a silvery color in late summer, and become golden brown over winter. Toes on hind flippers are all similar in size. Congregate on the shores of the Pribilof, Aleutian, and Channel Islands to breed and pup, then scatter to feed at sea.

GUADALUPE FUR SEAL *Arctocephalus townsendi* ♂ 1.9–2.4m, 150–220kg; ♀ 1.4–1.9m, 40–55kg

Males have a large head with a long pointed snout. Females are like Northern Fur Seals, but have larger flippers, with fur on the foreflipper. Toes on hind flippers are all similar in size. Males do not have enlarged head crest found in other fur seals. Dry fur is brown or dusky black and has a thick, grizzled appearance compared with coarser hair of sea lions. Now breeds and pups only at Guadalupe and Islas San Benito in Baja California, Mexico. A threatened species, rookeries on Southern California islands were hunted to local extinction at the turn of the nineteenth century. Uses precipitous rocky coasts and caves. Hunts fish and squid in the open sea.

STELLER'S SEA LION *Eumetopias jubatus* ♂ 2.7–3.2m, 500–1120kg; ♀ 1.9–2.9m, 263–365kg

Bearlike head with a short straight snout. Larger and paler than California Sea Lion. Males have long, coarse hair on massive chest, neck and shoulders. Pups are born with a dark brown fur that molts to a lighter color after three months. First toe on hind flipper is largest. Skull unique with conspicuous space between upper fourth and fifth post-canine teeth. An endangered species that is declining in numbers. Hauls out all along our west coast for pupping and breeding, most common in Alaska. Swims far from shore to fish.

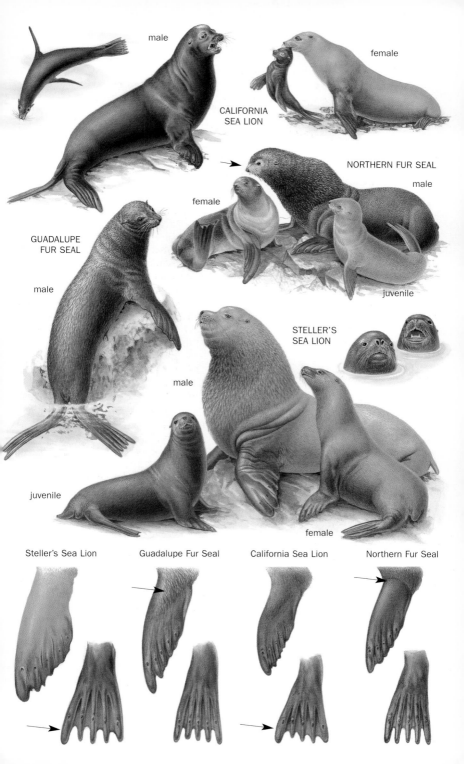

male

CALIFORNIA
SEA LION

female

NORTHERN FUR SEAL

male

GUADALUPE
FUR SEAL

female

juvenile

male

STELLER'S
SEA LION

male

juvenile

female

Steller's Sea Lion   Guadalupe Fur Seal   California Sea Lion   Northern Fur Seal

PLATE 87
PHOCID SEALS

PHOCID SEALS – These earless seals are awkward on land and cannot stand on their hind flippers. However, they are very specialized for deep, lengthy underwater dives.

HARP SEAL *Phoca groenlandica* 1.7–1.9m, 115–140kg

A white seal with a black lyre-shaped marking on back. Face is black. Males are slightly larger. Pups are white and turn silver-gray with black blotches as juveniles. Some females retain silver coat for up to eight years. Some males develop a dark "sooty" coat. Swim in large herds throughout the North Atlantic. Thousands haul out together on the pack ice to pup and nurse young.

RINGED SEAL *Phoca hispida* 1.0–1.5m, 45–107kg

A dark gray-colored seal with light rings on the body. Males are slightly larger. Pups are white; juveniles are silvery in color without rings. Much smaller than the Harbor Seal, with almost no neck and a short face. Strong nails on front flipper used to carve ice lair under snow drifts where they hide from predators and the weather. Females give birth in these lairs; males maintain their own lairs. Some populations are vulnerable to overhunting, but these seals are generally abundant on and around pack and fast ice.

RIBBON SEAL *Phoca fasciata* 1.5–1.6m, 70–80kg

A brown seal with whitish ribbons around their head and flippers. Male is dark brown; female is lighter. The pup's white coat is shed at five weeks for a gray coat that is bluish above and silvery below. More slender than other seals, with a long neck and flippers. Pups on heavy pack ice, and moves to the open sea in late spring and summer.

HOODED SEAL *Cystophora cristata* ♂ 2.3–2.8m, 200–435kg; ♀ 2.0–2.3m, 150–350kg

Silvery-gray seals with black splotches and conspicuous ornaments in the males. Female is distinguished from splotchy Gray Seal by having a shorter snout. Male is larger, and has a dark hood on the top of its head that can be inflated. Additionally, it can inflate its nasal septum like a red balloon. Both tricks probably attract females and intimidate rival males. Pups are silvery blue above and white below. Comes to pack ice around Labrador and Davis Strait to give birth, then migrates to areas around Greenland.

HARP SEAL

female

sooty
male

normal
male

pup

juvenile

ice lair

RINGED
SEAL

male

female

juvenile

RIBBON
SEAL

male

female

HOODED SEAL

female

pup

males: normal   extruded nasal septum

inflated hood

PLATE 88
GRAYISH SEALS

SPOTTED SEAL *Phoca largha* 1.4–1.7m, 81–109kg

A grayish seal with dark splotches, typically found on ice around Alaska. Overlaps little with the very similar Harbor Seal. Distinguished from Harbor Seal by range, having white (not dark) pups and a smaller, more delicate skull. Eats fish, shrimp, cephalopods, and crustaceans along coastal waters. Hauls out on ice.

HARBOR SEAL *Phoca vitulina* ♂ 1.4–1.9m, 75–150kg; ♀ 1.2–1.7m, 60–110kg

A wide-ranging grayish seal usually found hauled out on rocks. Color pattern may be dark with irregular pale rings or pale with dark splotches. Pups typically molt *in utero* and are born with a dark pelage. Often has banana-shaped profile when on rocks. Typically shy when hauled out, but sometimes habituate to humans. Most populations are recovering with recent protection from hunting. Common on undisturbed beaches, ledges, and rocks.

GRAY SEAL *Halichoerus grypus* ♂ 2.0–2.7m, 240–320kg; ♀ 1.6–2.2m, 150–260kg

A large gray seal with an exaggerated snout. Larger than the Harbor Seal, with a less doglike face. When viewed head-on the nostrils are curved resembling the letter W (not heart shaped like in the Harbor Seal). The color of males varies from black to gray-green, and can be solid or mottled. Females are typically silvery with dark patches, rarely solid black or cream-colored. Males are larger, with longer, broader snouts and more massive necks that are often scarred from fighting. Pups are born with white fur that molts to adult color at two to four weeks. Breeds and molts on land or ice, feeds on fish in the open sea.

CARIBBEAN MONK SEAL *Monachus tropicalis* 2.1–2.4m, 70–140kg

An extinct brownish seal from the Caribbean. Adults were pale brown on the back and lighter on the belly. This was the only seal from the Gulf of Mexico, and was recorded off the Florida and Texas coast. It preferred to haul out on sandy beaches, where they had little fear of humans, who hunted them to extinction. The last confirmed sighting was 1952. Its two sister species, the Hawaiian and Mediterranean Monk Seals, are both seriously Endangered. Juvenile Hooded Seals sometimes wander south into the Caribbean, and can be distinguished by having a much shorter face and by being smaller.

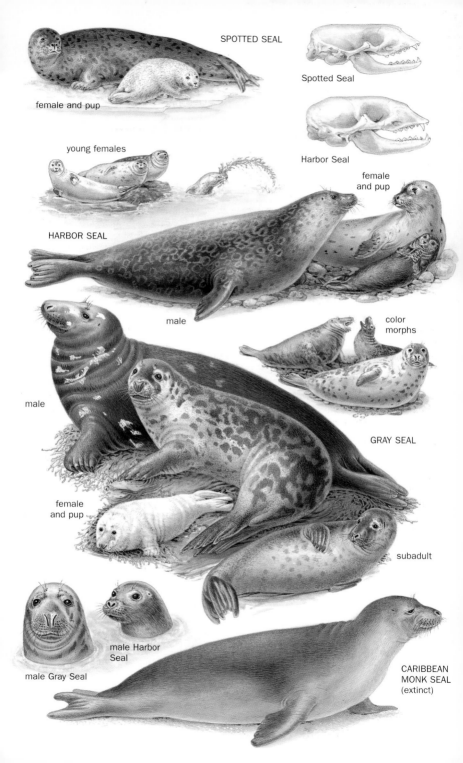

SPOTTED SEAL

female and pup

Spotted Seal

Harbor Seal

young females

HARBOR SEAL

female and pup

male

color morphs

male

GRAY SEAL

female and pup

subadult

male Gray Seal

male Harbor Seal

CARIBBEAN MONK SEAL (extinct)

PLATE 89
BIG SEALS AND MANATEE

WEST INDIAN MANATEE *Trichechus manatus* 2.7–3.5m, 500–1650kg

Unmistakable slow-moving mammal with a blunt nose and a broad spatulate tail. Gray skin is often green from algae growing on back. Young nurse from a nipple under the flipper. Endangered and vulnerable to collisions with speed boats. Feeds on marine plants in shallow Florida waters. Occasionally wanders as far north as the Chesapeake Bay.

WALRUS *Odobenus rosmarus* ♂ 2.5–3.5m, 590–1656kg; ♀ 2.3–3.1m, 400–1250kg

A large pinkish seal with tusks. Broad snout is covered with short whiskers. The color of an individual can vary from white to pink to reddish brown as blood flow to the skin changes for temperature control. Atlantic Walrus is smaller, with smoother skin and more rounded snouts than Pacific Walrus. Male is larger, with larger tusks and thick skin on the neck and shoulders. Tusks often break in fights between males. A clumsy walker on ice and rocks, it is an adept diver to forage on prey dwelling on the ocean bottom.

NORTHERN ELEPHANT SEAL *Mirounga angustirostris* ♂ 3.6–4.2m, 1500–2300kg; ♀ 2.2–3.0m, 400–800kg

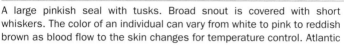

Large seal with a uniform brownish back and yellowish belly. Male is unmistakable with its enormous size, elephantine nose, and thickened neck. Canines of males are often exposed, and used to scar the necks of their male competitors. Newborns are black, and molt to a silvery color after weaning. Populations are recovering from historic overhunting. Hauls out on beaches to breed and molt, then migrates into the open ocean to feed on fish and squid.

BEARDED SEAL *Erignathus barbatus* 2.0–2.6m, 225–360kg

A large seal with a small head and prominent whiskers. Foreflippers are square-shaped. Whiskers are straight when wet and curved when dry. Female may be slightly longer than male. Adults are gray or brownish in color, sometimes with red on the face or flippers. Newborns are dark-colored with white patches on head and back. Juveniles often have irregular splotches of color on their heads and bodies. Feeds on prey caught at the sea bottom. Lives at low densities on moving ice and in open water.

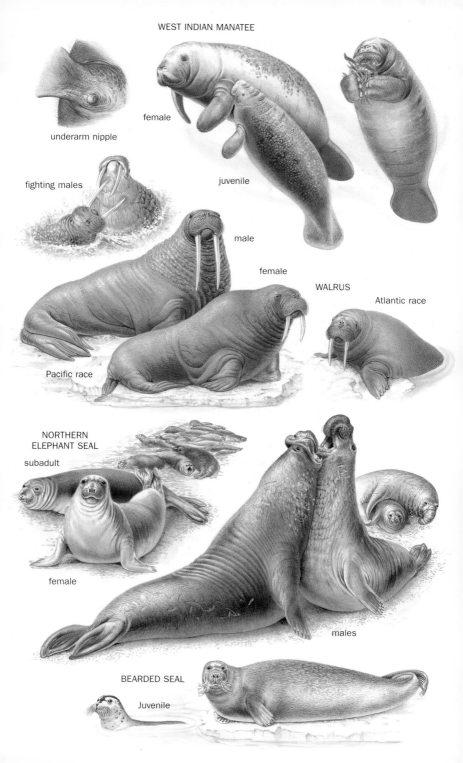

WEST INDIAN MANATEE

underarm nipple

female

juvenile

fighting males

male

female

WALRUS

Atlantic race

Pacific race

NORTHERN
ELEPHANT SEAL

subadult

female

males

BEARDED SEAL

Juvenile

PLATE 90
ANTELOPES AND PIGS

PRONGHORN *Antilocapra americana* ♂ 1.3–1.4m, 10–15cm, 42–59kg; ♀ 1.3–1.5m, 10–13cm, 41–50kg

North America's only antelope has unique coloration and horns. Has a stocky body, long legs, and short black horns. Cinnamon-colored body, with a white rump, belly, and facial markings. Male horns have a forward point-ing "prong," while female horns are smaller, usually lacking prongs. The horn sheaths are shed and quickly regrown each year. Some females do not grow the sheaths. Male is larger and has a black line on the lower jaw. Hairs on the rump and back of the neck can be erect-ed. Heavy eyelashes serve as sun shades. Feeds on a variety of plants, especially forbs and shrubs. Relies on good vision and speed to escape predators in open, grassland habitats.

BLACKBUCK *Antilope cervicapra* 1.2–2.1m, 10–17 cm, ♂ 20–57kg; ♀ 20–33kg

An exotic, bicolored antelope introduced in Texas. Adult males have long, V-shaped, spiral-ing horns and are black above. Females and young males are tan above and usually do not have horns. Males lighten in color after the spring molt. Coloration is accentuated by white eye rings, chin patch, chest, belly, and inner legs. Native to India and Pakistan. Introduced widely in Texas, with more than 7000 animals counted in 51 counties in 1974, mostly on the Edward Plateau. Does not survive long periods of freezing temperatures. Grazes on short to mid-length grasses and some brush.

COLLARED PECCARY *Pecari tajacu* 85–102cm, 3–5cm, 15–25kg

A native piglike animal with grizzled dark coat and a whitish collar. The hairs on the head and back can be erected into a mane. Small, incon-spicuous tail. This neotropical animal is restricted in the north by cold. Pro-duces a variety of grunts, snorts, and clicks. Herds of 15–20 animals root and browse in grasslands, desert scrub, and arid woodlands.

WILD BOAR *Sus scrofa* 1.3–1.8m, 15–30 cm, 35–200kg

Basically a domestic pig gone wild, typically covered with thin, coarse grizzled black and gray hair. Hybrids may be variable in color, including spotted, black and tan. A mane of long bris-tles may develop down the back. The upper teeth (tusks) curve upward and are often con-spicuous. Old World pig species widely introduced in North America. Feral populations are known from at least 18 states, especially in the humid south. Prolonged winter frost appears to restrict its winter foraging and, therefore, prevent its spread northward. Its rooting can be quite damaging to native vegetation. Shy and intelligent, difficult to see in their forest haunts.

PRONGHORN

male

female

horn
sheath

bony
core

horn anatomy

group at waterhole

BLACKBUCK

male

COLLARED PECCARY

WILD BOAR

Wild Boar wallow

PLATE 91
GOATS AND SHEEP

MOUNTAIN GOAT *Oreamnos americanus* ♂ 1.2–1.8m, 8–20cm, 46–136kg; ♀ 1.3–1.5m, 8–14cm, 45–83kg

A white goat with small, slightly curved black horns. White coat is long and shaggy in the winter and shorter in the summer. Males are larger, with longer horns. Similar to female Dall's Sheep, but with black, not brown horns. Typically migrates down slope in the winter and uphill in the summer. Uses open country above the timberline, with meadows for foraging and steep rocky cliffs or talus slopes for refuge from predators.

BIGHORN SHEEP *Ovis canadensis* ♂ 1.6–1.9m, 8–12cm, 75–135kg; ♀ 1.5–1.7m, 7–12cm, 48–85kg

A brown sheep with a white rump patch and large curved horns. Large male horns are used in combat to establish dominance. Female horns are smaller. Easily distinguished from Dall's Sheep by color and geographic range. In spring molt, cream-colored females are distinguished from Mountain Goats by coat and horn color. Heaviest in October, lightest in May. Measurements given are for *O. c. canadensis* from the central and northern Rockies, the largest subspecies. Often seen at saltlicks. Typically in groups of 5–80 animals. Many populations are migratory. Limited by water availability in desert habitats, where many isolated populations are declining and the subspecies *O. c. californiana* from California is endangered. Prefers treeless areas with nearby cliffs or rocky areas to escape from predators.

DALL'S SHEEP *Ovis dalli* ♂ 1.3–1.8m, 7–12cm, 73–110kg; ♀ 1.3–1.6m, 7–9cm, 46–50kg

A northern sheep with large horns. Coat color is usually uniform white; the subspecies *O. d. stonei* from Yukon and British Columbia is silvery gray with a white muzzle, leg trimmings, and rump patch. Male is larger than female, with larger horns. Horns are a bit smaller than in the Bighorn Sheep, and more widely flaring. Winter coat is much thicker. Feeds on grasses and low shrubs. Typically migrates between summer and winter ranges in its rugged mountain habitat.

MOUNTAIN GOAT

male

BIGHORN SHEEP

female

typical

DALL'S SHEEP

O. d. stonei

male

female

PLATE 92
BISON AND NILGAI

AMERICAN BISON *Bison bison* ♂ 3.0–3.8m, 43–90cm, 460–907kg; ♀ 2.1–3.2m, 43–90cm, 360–544kg

North America's largest land mammal. Has a distinctive humped profile with larger forequarters and smaller hindquarters; these traits are more pronounced in males. The brown wooly pelage is thickest around the neck, extending onto the shoulders and back in males. Males also have larger horns, which are stouter and more curled; female horns are more slender with upward pointing tips. Calves are reddish. Hair is longer in the winter. The Endangered Wood Bison (*B. b. athabascae*) from western Canada is slightly taller, darker, and woolier, with a larger hump than the Plains Bison (*B. b. bison*). Wallows 2–3m wide are used by all ages and sexes. Prairie rock rubs worn smooth by 10,000 years of use usually rest at the bottom of a slight depression formed by countless hoofs circling the stone as the bison rubbed against it. Nearly hunted to extinction (probably fewer than 1000 animals left) at the end of the nineteenth century, the species had recovered to ca. 150,000 animals by 1995, with 90% privately owned. Typically grazes in herds of 4–20, these groups sometimes merge into much larger congregations. Reaches its highest density in mixed and short-grass prairies, but can survive in a variety of open habitats.

NILGAI *Boselaphus tragocamelus* 2.2–2.4m, 40–45cm, ♂ 109–306kg; ♀ 109–213kg

A very large, brownish exotic antelope introduced and running wild in Texas, especially south in Kennedy and Willacy counties. This large animal stands up to 1.5m at its enlarged shoulders; its profile then slopes to the smaller rump. Generally brownish, the bull is slightly gray while the female and young are orangish brown. It has white on the face, chin, and throat. Below this white bib hangs a beard of hair, which is larger in males. Female is smaller and generally lacks horns. Typically herds in groups of 10. Intolerant of cold temperatures. Grazes and browses in relatively dry areas of flat to rolling country with a moderate cover of thin forest or scrub. Avoids heavy woods.

female

male

Plain's Bison

AMERICAN BISON

wallow

Wood Bison

NILGAI, male

PLATE 93
ARCTIC UNGULATES

CARIBOU *Rangifer tarandus* ♂ 1.6–2.1m, 11–22cm, 81–153kg;
♀ 1.4–1.9m, 10–20cm, 63–94kg

A stocky deer of the far north. Both sexes have antlers; male antlers are larger and semi-palmated, especially the single, flat brow tine that extends down almost, but not quite, past the nose; female antlers are much smaller and less palmate. Female tends to have more white hairs than bulls. Old antlers are whitish, new ones are black from their velvet covering. Coloration and antler size vary across subspecies. The Woodland Caribou from boreal forests and alpine tundra is the largest caribou, and is brown in the summer and grayish in the winter. It has creamy white hair on the neck, mane, underbelly, rump, and on a patch above each hoof. Barren Ground Caribou uses taiga forests and tundra and is medium in size; the coat is chocolate brown in summer, lighter brown in winter. Pearly Caribou from high Arctic islands is smaller with shorter legs, face, and ears, and a lighter coloration. Caribou eat grasses and shrubbery leaves in the summer, lichens in the winter. Large migratory herds calve on the tundra in the spring, then wander the tundra and forest searching for food. Forest and mountain Caribou migrate less. The Woodland Caribou subspecies (*R. t. caribou*) is Endangered. Domesticated and European Caribou are known as Reindeer. Depending on the subspecies and time of year, the Caribou uses a variety of boreal forest and treeless tundra and mountain habitats.

MUSKOX *Ovibos moschatus* ♂ 2.1–2.6m, 7–12cm, 186–410kg;
♀ 1.9–2.4m, 6–12cm, 160–191kg

A stocky, shaggy bovid of the arctic. The long, brown winter hairs extend nearly to the ground, covering fine, cashmere-like underhairs. Stockings and saddle are creamy white. Male is larger (some reach 650kg in captivity) and has larger horns that merge into a massive boss on the forehead. Female horns are shorter, more slender, and more curved. Feeds on sedges, grasses, and willows. Typically lives in groups, although some males are solitary in summer. When confronted by would-be predators, adults circle around calves confronting the enemy with a ring of pointy horns. Nearly extinct at the end of the nineteenth century, now protected, reestablished in Alaska and recovering in Canada. Uses Arctic tundra, preferring grassy areas in the summer and windswept areas with exposed vegetation in the winter.

Barren
Ground
Caribou

female

male

CARIBOU

male
Woodland
Caribou

male Pearly Caribou
with antlers in velvet

defensive circle

MUSKOX

female

male

PLATE 94
ELK AND MOOSE

ELK *Cervus elaphus* ♂ 2.1–2.6m, 11–17cm, 178–479kg; ♀ 2.0–2.5m, 8–19cm, 171–292kg

A large, tan-colored ungulate with a darker neck and white rump. In season, the male has a shaggy neck mane and enormous antlers consisting of one main beam and, typically, six points. Antlers are usually shed in February and regrow over the summer. Summer coat is sleek and tawny brown; winter coat is grayish brown. Fawns are spotted. Tule Elk (*C. e. nannodes*) from parts of California are lighter over-all in color and smaller. The Elk is smaller and paler than the Moose, without palmate antlers; larger than deer, with unique dark neck/white rump coloration. During the fall rut, males repeatedly give a high-pitched "bugle" vocalization. Lives in herds sometimes exceeding 200 in open habitats, smaller groups in forested areas. Hunted to extinction throughout eastern states by the mid 1800s; recent reintroductions of small populations into Michigan, Kentucky, Tennessee, and Pennsylvania are encouraging. The smaller *C. elaphus* subspecies from Europe is called the Red Deer and typically has rougher antlers that grow up and inward rather than up, out and backward. Browses and grazes on a variety of plant species, prefer-ring open or brushy habitats to mature forest.

MOOSE *Alces alces* ♂ 2.4–3.1m, 8–12cm, 360–600kg; ♀ 2.3–3.0m, 8–12cm, 270–400kg

Largest deer with unique hanging dewlap. Elongated head with pendulous snout and huge rotatable ears. Upper lip marked by small rectangular bare spot. Long legs gray at bottom. Heavy body with humped shoulders and a small tail. Light brown to black body color from guard hairs over gray undercoat. Male is larg-er than the female with enormous palmate antlers that are grown in summer and shed in winter. Most females have white hair around vulva. Juvenile pelage is reddish. Mostly soli-tary; herbivorous; active any time, especially dusk and dawn. Vocalizations in the early fall rut include deep grunts and moos. Mud wallows marked by large tracks; most antler rubs on vegetation are higher (100–200cm) than deer (less than 115cm). Recently reintroduced in Colorado and northern Michigan. Often unwary but can be dangerous up-close; abundant in northern boreal forests, especially wet areas.

male

female

ELK

MOOSE

wallow

antler
rub

male

female

juvenile

PLATE 95
NATIVE DEER

DEER – These two deer species are among our most familiar animals. They are distinguished by differences in the ears, facial coloration, tail, and antlers. Hybrids are rare, and typically sterile. In areas of overlap, the Mule Deer inhabits drier habitats. The White-tailed Deer sticks to moister habitats, and is expanding its range westward. Both leave their sign on the landscape in similar ways. Antlers are rubbed on the stems of saplings, scraping away the bark. Bark scraped off with their incisors leaves a different, chewed pattern on small trees. Chewed branch ends are cut off roughly, leaving a frayed end, typically around human knee- or hip-height. Matted grass beds mark their resting sites. Finally, males mark their areas with dirt scrapes dug by their hooves and flagged with broken branches above and urine scent in the mud.

MULE DEER *Odocoileus hemionus* ♂ 1.3–1.7m, 13–22cm, 40–120kg; ♀ 1.2–1.6m, 12–21cm, 30–80kg

Ears are larger and tail is smaller than that of the White-tailed Deer. Males have dichotomously branching antlers that are usually shed in January and regrow over the summer. The tail is large and black in coastal subspecies, smaller with a black tip in interior subspecies. Interior animals are pale brown or tan in the winter with a large white rump patch, while coastal animals are darker and grayer with a smaller white rump patch. All races are rusty red in new summer coat. Fawns are reddish with white spots. There is a V-shaped dark mark from the point between the eyes, upward and laterally, especially in males. When alarmed they bound away in a "stot," with four feet hitting the ground together at each bound (the White-tailed Deer springs from hind to front feet). Populations in northern mountains migrate up to higher elevations in the summer and down to the foothills in the winter. Prefers mixed habitat with both open areas for feeding and forest or brushy areas for protection. Common in western mountain forests, deserts, and brushlands.

WHITE-TAILED DEER *Odocoileus virginianus* 0.8–2.4m, 10–37cm, ♂ 22–137kg; ♀ 30–90kg

Ears are smaller and tail is larger and whiter than in the Mule Deer. Males have antlers consisting of smaller vertical tines branching off the single main beam. Year-old male fawns have small "buttons" of antlers. Antlers are shed in December and January and regrown over the summer. Long tail is brown above, white below, and fringed in white on the sides. Coat is reddish brown to bright tan in the summer, duller and grayer in the winter. Fawns are reddish and spotted with white. Northern populations are larger; the Endangered Dwarf Key Deer (*O. v. clavium*) from the Florida Keys stands 60cm at the shoulder and weighs ca. 35kg. The Columbian White-tailed Deer subspecies (*O. v. leucurus*) from coastal Oregon and Washington is also Endangered. Prefers forest edges and open woodlands near brushlands, especially old fields and agricultural areas. Uses a variety of forested habitats from temperate to tropical, semiarid to rain forest.

male in velvet

coastal race
summer coat

inland race
winter coat

female

MULE DEER

antler
rub

incisor
scrape

grass bed

dirt scrape

WHITE-TAILED
DEER

male
winter coat

fawn

female
summer coat

PLATE 96
EXOTIC UNGULATES

COMMON FALLOW DEER *Dama dama* 1.4–2.0m, 15–23cm, ♂ 79–102kg; ♀ 36–41kg

An exotic spotted deer with palmate antlers. Originally from the Mediterranean, escapes and releases from game farms have started free-ranging populations in at least seven states and provinces (California, Georgia, Texas, Alabama, Kentucky, New Mexico, and New Brunswick) and islands off British Columbia; many less successful introductions have also occurred. Most animals are rust or tan-colored, with white rump patch and belly; winter pelage is darker and the spots are often indiscernible. The white spots on the back merge into a white line near the rump and there is a black or tan line on the back. Melanistic, very dark, and very light morphs are also known, and may predominate in domesticated herds. Prefers mixed habitats with open areas for feeding and covered areas for winter food and shelter.

CHITAL *Axis axis* 1.0–1.75m, 12–38cm, ♂ 30–75 kg; ♀ 25–45kg

A spotted exotic deer with three tines on each antler. This brownish deer has many delicate white spots, and a white abdomen, rump, throat, insides of legs and ears, and underside of tail. A dark stripe runs down the center of the back. Originally from India, Nepal, and Sri Lanka; they are now the most abundant exotic ungulate in Texas. Many have escaped from game farms in other states and probably also survive in the wild in California and Florida. Prefers secondary forests mixed with grassy areas.

SIKA DEER *Cervus nippon* 1.0–1.5m, 7–13cm, ♂ 68–109kg; ♀ 45–50kg

A small deer from east Asia with introduced feral populations surviving in Texas and Maryland. This compact deer has a distinctive wedge-shaped head. Dark brown with a variable amount of white spotting that is absent in some animals. Texas populations are especially variable due to extensive hybridization. There is a white rump patch that is most visible when the animal is alerted. The male's antlers have three or four points branching off each main beam; the female has a corresponding pair of black bumps on the forehead. Browses and grazes in a variety of woodlands.

SAMBAR *Cervus unicolor* 1.9–2.8m, 25–30cm, 109–260kg

A large dark brown ungulate with stout antlers and a white rump and undertail. Has a bald, rounded glandular area on the middle of the throat. Large antlers are stout and rugose, with three prominent tines. In season, males have a mane of hair on the neck and forequarters. Crepuscular and nocturnal. Native to India and Southeast Asia, now established in small numbers in Texas, California (Obispo County), Florida (St. Vincent Island), and perhaps other areas. Prefers wooded areas.

BARBARY SHEEP *Ammotragus lervia* 1.4–1.8m, 14–25 cm, ♂ 100–145kg; ♀ 45–65kg

A large exotic sheep with long, curved horns and long hair on the throat, chest, and front of the legs. Native to the Sahara Desert of North Africa, introduced into California, New Mexico, and Texas (mostly Panhandle region). Tail is long and tufted. Upperparts are rufous or grayish brown with a blackish mid-dorsal line. Flanks, inner surface of legs and belly are whitish; chest is colored like the sides. Yellowish horns darken with age. Inhabits dry, rough, barren, and waterless habitat, likely competing with native Bighorn Sheep.

male

COMMON
FALLOW DEER

female

male

CHITAL

male

SIKA DEER

female

male

SAMBAR

male

female

BARBARY SHEEP

PLATE 97
LARGE WHALES
WITHOUT DORSAL FINS

GRAY WHALE *Eschrichtius robustus* ♂ 11.1–14.3m; ♀ 11.7–15.0m, 15,700–33,800kg

Unique with mottled gray color and low rounded hump in place of dorsal fin. Narrow triangular head is covered with barnacles and whale lice. Mouthline is straight or slightly arched. Small flippers and large flukes are frequently marked with white from lice and scars. Top of tail stock is covered with bumpy knuckles. Active in water, often swims in shallows. Stirs up mud when feeding from sea floor. Strains food with baleen. Low heart- or V-shaped blow. Migrates along west coast between Arctic summer waters and Baja winter waters.

BOWHEAD *Balaena mysticetus* ♂ 14–17m; ♀ 16–18m, 7500–10,000kg

Large, smooth-skinned, black whale with no dorsal fin. Indentation behind blowhole divides profile of triangular head from rotund body. Head and back produce two humps above water in profile. Chin is white with variable black spotting; some also have white around tail stock. Large fluke and paddle-shaped flippers are black and unmarked; fluke is shaped like that of the North Atlantic Right Whale. A record of 3m-long Bowhead baleen is largest of any whale. Blow is high and V-shaped. Feeds alone or in groups. Can break ice to breathe. A rare and Endangered species. Closely tied to pack ice in Arctic waters.

NORTH ATLANTIC RIGHT WHALE *Eubalaena glacialis* 11–18m, 25,000–72,000kg

Large black whale with large callosities and no dorsal fin. Smooth slope of back is different from Bowhead's double-hump profile. Flukes and wedge-shaped flippers are black with limited mottling. Body is black except for variable amounts of white on belly. Blow is V-shaped. This slow, inquisitive whale was the "right" whale to hunt and was nearly driven to extinction. Modern recovery is slowed by mortality from fishing gear entanglement; still Endangered. Rare but approachable near shore on both coasts.

SPERM WHALE *Physeter catodon* ♂ 11–18.3m, 11,000–57,000kg; ♀ 8.3–12.5m, 6800–24,000kg.

Unique large whale with square head, undersized lower jaw, and wrinkled skin. Body color varies from dark gray to light brown, rarely white like Moby Dick. Skin around mouth usually white. Male is larger, with a protruding spermaceti organ projecting beyond the skull. Angular hump is followed by a keeled tail stock topped by bumpy knuckles. Has stubby flippers and a large triangular fluke. Single blowhole at front of head sends a bushy blow forward and to the left. Typically dives to 300m in search of squid and octopi, may reach 3000m. Usually dives to avoid boats. An endangered species. Swims in groups of ca. 20 animals in deep waters; moves north in summer and south in winter.

spy hopping

close up of Whale
Louse and barnacle

GRAY WHALE

bottom feeding

BOWHEAD

North Atlantic
Right Whale fluke

NORTH ATLANTIC
RIGHT WHALE

calf

feeding at surface

male

female

SPERM WHALE

PLATE 98
ENORMOUS WHALES WITH DORSAL FINS

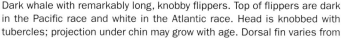

HUMPBACK WHALE *Megaptera novaeangliae* 14–17m, 25,000–45,000kg

Dark whale with remarkably long, knobby flippers. Top of flippers are dark in the Pacific race and white in the Atlantic race. Head is knobbed with tubercles; projection under chin may grow with age. Dorsal fin varies from low and stubby to high and curved. Pattern of white on flukes allows identification of individual whales; end of tail is serrated. Widely spaced throat grooves help distend mouth for filter feeding. Blow is low and bushy. Most active and acrobatic of the large whales. This Endangered species migrates along both coasts between summer Arctic waters and winter southern waters.

BLUE WHALE *Balaenoptera musculus* 22–28m, 64,000–195,000kg

The world's largest animal, mottled blue with a U-shaped head and tiny dorsal fin. Larger than Fin Whale, with a blue mottled back (not smooth gray), a more rounded head, and a less prominent dorsal fin. Shape of dorsal fin varies from round to pointed. Belly may be light blue, white, or yellow. Long slender flippers tipped in white. Thick, smooth tail stock. Females are larger than males. Animals are larger in the southern hemisphere. High, slender blow. A rare and Endangered species. Migrates along both coasts between summer Arctic waters and winter southern waters.

FIN WHALE *Balaenoptera physalus* ♂ 17.7–22m; ♀ 18.3–24m, 45,000–70,000kg

Large whale with smooth gray skin and a V-shaped head. Undersides of belly, flippers, and fluke are white. Rarely shows fluke. Shape of pointed dorsal fin is variable. Top of tail stock is ridged. Asymmetric lower jaw and baleen are dark on left and white on right. This head color, and the unique chevron pattern behind the head allow individual identification. Blow is high and cone-shaped. A rare and Endangered species. Typically swims in deeper waters along both coasts.

Humpback
blow

Blue
blow

HUMPBACK
WHALE

Fin blow

BLUE WHALE

newborn

FIN WHALE

PLATE 99
LARGE WHALES
WITH DORSAL FINS

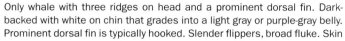

BRYDE'S WHALE *Balaenoptera edeni* ♂ 11.9–14.8m; ♀ 12.2–15.6m, 11,300–16,200kg

Only whale with three ridges on head and a prominent dorsal fin. Dark-backed with white on chin that grades into a light gray or purple-gray belly. Prominent dorsal fin is typically hooked. Slender flippers, broad fluke. Skin may be mottled with scars. Baleen is black with a pale gray, coarse fringe. Blow is moderately high and thin. Spectacular breaches often repeated two or three times. Uses warm waters, typically south of Chesapeake Bay in the east and south of Baja in the west.

SEI WHALE *Balaenoptera borealis* 14–18.6m, 15,000–40,000kg

Large whale with one central ridge on head and a prominent dorsal fin. Dorsal fin is less hooked than that of Bryde's Whale, and baleen is white and silky. Has thin, pointy flippers and a triangular fluke. Back is blueish gray, belly and chin are white. Skin may be mottled with scars. Blow is low and forked. A rare and endangered species. Rarely breaches. Found on both coasts, far from shore.

NORTHERN MINKE WHALE *Balaenoptera acutorostrata* ♂ 7.6–9.8m; ♀ 7.3–10.7m, 5000–13,000kg

Unique with white band on flippers and a sharply pointed snout. Flippers are rarely all dark. Hooked dorsal fin is largest, in relative size, of all baleen whales. Dark gray back. White on belly and chin extends dorsally in center of back. Baleen is yellowish white. Rarely has visible blow. Common, especially in cooler waters. Often seen near coast.

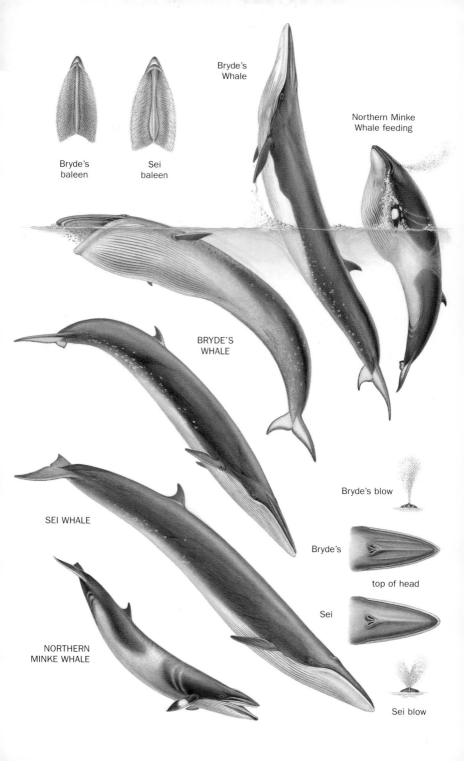

Bryde's
baleen

Sei
baleen

Bryde's
Whale

Northern Minke
Whale feeding

BRYDE'S
WHALE

Bryde's blow

SEI WHALE

Bryde's

top of head

Sei

NORTHERN
MINKE WHALE

Sei blow

PLATE 100
SMALL WHITE AND GRAY WHALES

SMALL SPERM WHALES – These are two of the smallest whales. Both have blunt, squar-ish heads and a conspicuous white crescent behind the eye forming a false gill. Their lower jaw is small and underslung. They are sharklike in appearance, with a dark bluish-gray back blending into a light gray or pink belly. The two species are distinguished by body size, posi-tion and size of the dorsal fin, and tooth number. Shy, sometimes leave a reddish-brown cloud of fecal "ink" if startled. They live in deep waters and are difficult to see as they typ-ically surface and dive quietly.

DWARF SPERM WHALE *Kogia simus* 2.1–2.7m, 136–272kg

Has a prominent dorsal fin in center of back. Smaller than the Pygmy Sperm Whale, with fewer teeth (7–12 pairs). Seen off southeast coast, rare strandings along west coast.

PYGMY SPERM WHALE *Kogia breviceps* 2.7–3.4m, 318–408kg

Small dorsal fin is positioned one third of the way to the tail. Larger than Dwarf Sperm Whale, with more teeth (12–16 pairs). Sometimes has pale circular mark in front of eye. Occurs along both coasts, most often seen off the southeastern coast.

RISSO'S DOLPHIN *Grampus griseus* 2.8–3.8m, 400–600kg

Unique with very large dorsal fin and rounded head. Blunt melon is creased down the middle when viewed head-on. Body is typically gray and scarred, but may be dark gray to white. Young are an unscarred brownish gray. Scars accumulate with age, and body color lightens. Typically in groups of 3–50 animals in offshore waters along both coasts.

NARWHAL *Monodon monoceros* ♂ (without tusk) 4.8m, 1580kg; ♀ 4.1m, 960kg

Only whale with a tusk; mottled gray color and rounded head also unique. Typically, only males have a tusk, which is a hollow, spiraling, modified tooth. Rarely two tusks are present. Older animals are almost white in col-oration, and have a more convex trailing edge of their fluke. Tusks often break in fights over females, and scar the heads of males. Only found in high Arctic seas near pack ice.

BELUGA *Delphinapterus leucas* ♂ 3.5–4.9m, 800–1500kg; ♀ 3.3–4.0m, 540–790kg

White color and rounded head unmistakable. No dorsal fin, small broad flip-pers. Young are born slate gray to pinkish brown and mature into blueish-gray subadults. Skin turns white after sexual maturity, and may look yellow in some light. Males are larger, with a more pronounced melon. Chirps and whistles may be heard above the surface. Lives in groups of 5–20 along northern shorelines and estuaries.

DWARF
SPERM WHALE

PYGMY
SPERM WHALE

RISSO'S
DOLPHIN

NARWHAL

female

jousting male
Narwhals

BELUGA

female

subadult

PLATE 101
BLACK DOLPHINS

PILOT WHALES – These two bulbous-headed dolphins are jet black with a variable blaze of gray or white behind the eye, and light-colored patches on the belly and throat. They may also have a saddle of paler color behind dorsal fin. The forward-sitting dorsal fin is bulbous and hooked in males, more upright in females. Their flukes have conspicuously pointed tips. Strong blow often visible. In addition to range, they are distinguished by the size of their flippers and their teeth—both of which are difficult to measure at sea.

LONG-FINNED PILOT WHALE *Globicephala melas* ♂ 6.2m, 2320kg; ♀ 5.12m, 1320kg

Has long flippers with a sharp elbow-like bend. 8–12 pairs of teeth. Swims in groups of 10–50 in northern and temperate waters off the east coast.

SHORT-FINNED PILOT WHALE *Globicephala macrorhynchus* ♂ 5.5–7.0m; ♀ 4.25–5.0m, 900–3000kg

Short flippers have a gently curved edge. Seven to nine pairs of teeth. Nomadic groups of 10–30 in deep waters along the west, southeastern, and Caribbean coasts.

PYGMY KILLER WHALE *Feresa attenuata* 2.0–2.6m, 110–170kg

Small, black dolphin with rounded flipper tips and white "lips." Snout is more rounded than Melon-headed Whale. May have a subtle dark brown cape on back that does not dip into flanks. 8–13 pairs of teeth. Typically avoids ships. Found off southeastern coast and Caribbean.

MELON-HEADED WHALE *Peponocephala electra* 2.1–2.7m, 160kg

Small, black dolphin with pointed flipper tips and white or pink "lips." Snout is slightly more pointed than in the Pygmy Killer Whale. Back is dark, belly is lighter. May have a dark mask on face and a dark cape that dips toward belly on flanks. 20–25 pairs of teeth. Swims in groups of a few hundred far off the southeastern coast and Caribbean.

KILLER WHALE *Orcinus orca* 6.0–10.0m, 3500–7000kg

Large, black dolphin with high dorsal fin and white eyespot. Male is larger and has an erect dorsal fin, female has a smaller curved dorsal fin. Large paddle-shaped flippers grow with age. Belly, chin, and underfluke are white. Has variable white and gray swirls on sides. An acrobatic swimmer in all oceans including deep and shallow waters.

FALSE KILLER WHALE *Pseudorca crassidens* ♂ 3.7–6.0m; ♀ 3.3–5.1m, 1000–1360kg

Large, black dolphin with sickle-shaped flippers. Short flippers have a hump at the center and slightly concave tip. Chest may be gray. Dorsal fin may be pointed or round. Rare but widespread and approachable along both coasts.

female

male

LONG-FINNED PILOT WHALE

SHORT-FINNED PILOT WHALE

PYGMY KILLER WHALE

MELON-HEADED WHALE

KILLER
WHALE

male

female
and young

FALSE KILLER WHALE

PLATE 102
UNSTRIPED, BEAKED DOLPHINS

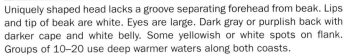

ROUGH-TOOTHED DOLPHIN *Steno bredanensis*
♂ 2.0–2.6m; ♀ 2.0–2.5m, 90–155kg

Uniquely shaped head lacks a groove separating forehead from beak. Lips and tip of beak are white. Eyes are large. Dark gray or purplish back with darker cape and white belly. Some yellowish or white spots on flank. Groups of 10–20 use deep warmer waters along both coasts.

BOTTLENOSE DOLPHIN *Tursiops truncatus* 2.6–3.4m, 150–300kg

Large, plain gray dolphin with long flippers. Variable in size and color, some populations may have a few spots. Body is darkest on back, grading to white on belly. Acrobatic and fond of bow-riding. Common in groups of 1–25 along coastlines and far offshore.

ATLANTIC SPOTTED DOLPHIN *Stenella frontalis*
1.7–2.3m, 100–143kg

Spotted dolphin with a gray band between eye and flipper. Dark and light color grade together on tail stock. Lips may be white. Calves born unspotted. Southern races are more heavily spotted than northern pelagic forms. Groups of 5–15 are mainly in warm Atlantic waters.

PANTROPICAL SPOTTED DOLPHIN *Stenella attenuata*
♂ 1.6–2.6m; ♀ 1.7–2.4m, 90–119kg

Spotted dolphin with prominent dark dorsal cape. Has dark gray band between jaw and flipper and a clear division of dark and light color on tail stock. Lips are white. Calves get first spots on belly. Dorsal fin shape is variable. Groups may reach thousands. Found in tropical waters and most of the Atlantic coast, where it is typically offshore.

NORTHERN RIGHT WHALE DOLPHIN *Lissodelphis borealis*
♂ 2–3.1m; ♀ 2–2.6mm, 60–113kg

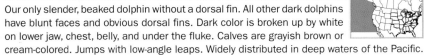

Our only slender, beaked dolphin without a dorsal fin. All other dark dolphins have blunt faces and obvious dorsal fins. Dark color is broken up by white on lower jaw, chest, belly, and under the fluke. Calves are grayish brown or cream-colored. Jumps with low-angle leaps. Widely distributed in deep waters of the Pacific.

high Bottlenosed leaps

ROUGH-TOOTHED
DOLPHIN

BOTTLENOSE DOLPHIN

ATLANTIC SPOTTED DOLPHIN

Carolinas south

Gulf Stream and
New England

PANTROPICAL
SPOTTED DOLPHIN

Spotted Dolphins

NORTHERN RIGHT WHALE DOLPHIN

PLATE 103
STRIPED, BEAKED DOLPHINS

SPINNER DOLPHIN *Stenella longirostris* 1.3–2.4m, 22–75kg

Characterized by spinning leaps, this species has a long, thin beak. The only other spinning dolphin, Clymene, has a shorter beak. Color may be dark "battleship" gray, or tricolored with a darker back and white belly. Some gray animals have white patches on belly. Stripe between eye and flipper is parallel (not triangular). Dorsal fin becomes more erect with age. Males have a large post-anal hump. Swims in groups of up to 200, sometimes with other species, in warmer Atlantic waters and south of Baja California in the Pacific.

CLYMENE DOLPHIN *Stenella clymene* 1.8–2.0m, 50–85kg

A tricolored spinning dolphin with a relatively short beak. Stripe between eye and flipper is triangular (not parallel). Often has white "moustache" on upper mandible. Tail stock of males is usually keeled. Smallish, curved dorsal fin is sometimes marked with pale coloration. Groups of up to 50, often mixed with other dolphin species. Found in warm Atlantic and Caribbean waters, where it is rarely seen.

STRIPED DOLPHIN *Stenella coeruleoalba* 8–2.5m, 110–156kg

Unique with dark eye-to-anus stripe and a pale marking below dorsal fin. Typically grayish, but may include blue or brown tones. Belly is white or pink. Specific stripe pattern is variable, but most originate from the eye. Groups of 10–500 use deep water along both coasts.

SHORT-BEAKED COMMON DOLPHIN *Delphinus delphis*
♂ 1.7–2.2m; ♀ 1.6–2.2m, 70–110kg

A short-beaked dolphin with yellow and gray hourglass coloration. The four-part hourglass color scheme includes dark back, white belly, yellowish flanks and a gray tail stock. Yellow may appear pale gray at sea. The variable flipper stripe typically originates with a zigzag near the middle of the lower mandible. Has a more rounded head and more contrasting colors than Long-beaked Common Dolphin, including a dark eye patch. There is considerable variability in the pattern of the colors and stripes. Common along both coasts.

LONG-BEAKED COMMON DOLPHIN *Delphinus capensis*
♂ 2–2.5m; ♀ 1.9–2.2m, 70–135kg

A long-beaked dolphin with hourglass coloration. Flipper stripe typically originates near the corner of the mouth. Has a more sloping head profile than *D. delphis*, and more muted colors. Groups of 10–500 found off southwestern coast.

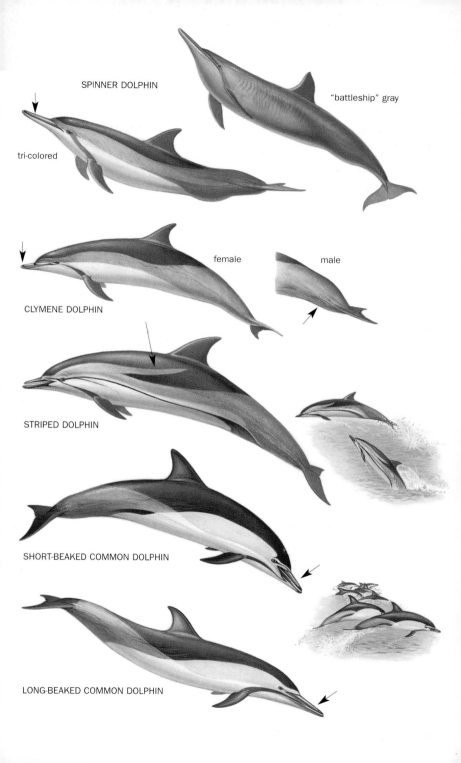

SPINNER DOLPHIN

"battleship" gray

tri-colored

CLYMENE DOLPHIN

female

male

STRIPED DOLPHIN

SHORT-BEAKED COMMON DOLPHIN

LONG-BEAKED COMMON DOLPHIN

PLATE 104
STRIPED, BLUNT-NOSED DOLPHINS AND PORPOISES

PACIFIC WHITE-SIDED DOLPHIN *Lagenorhynchus obliquidens*
1.7–2.5m, 75–200kg

Blunt-nosed dolphin with three-part color pattern and black beak. Back is dark gray, sides are streaked light and dark gray, belly and chin are white. Dorsal fin and flippers also colored with light and dark gray. Not shy of boats. Gregarious, active swimmer in offshore waters of the Pacific, sometimes associates with other dolphins or seals.

FRASER'S DOLPHIN *Lagenodelphis hosei*
♂ 2.3–2.7m; ♀ 2.1–2.6mm, 164–209kg

Short-beaked dolphin with a pair of broad, straight stripes running from eye to tail stock. Back is gray, belly is creamy white or pink. Thin, dark flipper stripe connects to beak. Dorsal fin and flippers are small. Sometimes associated with other dolphins. Rarely sighted off Caribbean coast.

ATLANTIC WHITE-SIDED DOLPHIN *Lagenorhynchus acutus*
♂ 2.3–2.8m; ♀ 1.9–2.4mm, 180–230kg

Unique with short beak and yellow patch on tail stock. Back is black, belly white, and flanks are gray with white and yellow patches. Thin black flipper-stripe starts from corner of mouth. Lower mandible is all white. Swims in groups of 5–50 in North Atlantic.

WHITE-BEAKED DOLPHIN *Lagenorhynchus albirostris*
♂ 2.5–3.1m; ♀ 1.8–2.4m, 180–354kg

Dark, robust-bodied dolphin with a white blaze on flanks and pale patch on tail stock. Head is dark gray, short beak and throat are white. Beak may be marked with brown or gray. Found in groups of up to 30 in North Atlantic, especially along edge of continental shelf.

PORPOISES – A group of small, blunt-nosed dolphins. Two species are known from our waters, mostly near the coast.

HARBOR PORPOISE *Phocoena phocoena* 1.5–1.9m, 45–90kg

Small gray porpoise with no beak. Lips and back are dark, throat and belly are white. Flippers are small and rounded. Wispy, dark flipper stripes. Shy and hard to watch. Swims in small groups of two to five near shore along western and northeastern coasts.

DALL'S PORPOISE *Phocoenoides dalli* 1.8–2.3m, 84–114kg

Unique blunt-faced porpoise boldly patterned with black and white. Fluke and dorsal fin also colored with black and white. Male is slightly larger and has a hump behind dorsal fin. Female has a black trident pattern in genital area. Active and unpredictable swimmer, frequently creates "rooster tail" splashes. Common in Pacific.

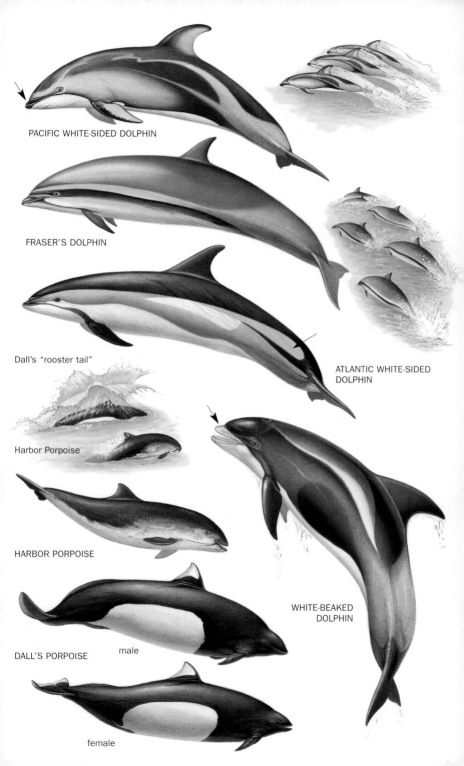

PACIFIC WHITE-SIDED DOLPHIN

FRASER'S DOLPHIN

Dall's "rooster tail"

ATLANTIC WHITE-SIDED DOLPHIN

Harbor Porpoise

HARBOR PORPOISE

WHITE-BEAKED DOLPHIN

DALL'S PORPOISE

male

female

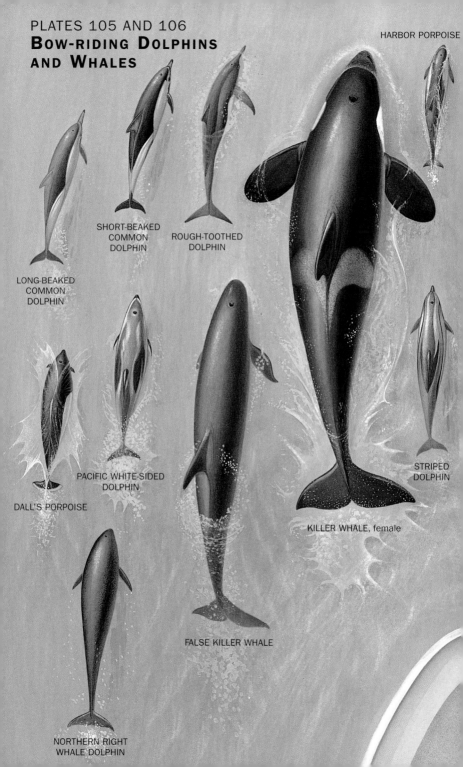

PLATES 105 AND 106
BOW-RIDING DOLPHINS AND WHALES

HARBOR PORPOISE

SHORT-BEAKED
COMMON
DOLPHIN

ROUGH-TOOTHED
DOLPHIN

LONG-BEAKED
COMMON
DOLPHIN

PACIFIC WHITE-SIDED
DOLPHIN

DALL'S PORPOISE

STRIPED
DOLPHIN

KILLER WHALE, female

FALSE KILLER WHALE

NORTHERN RIGHT
WHALE DOLPHIN